Dad,

Merry Christmas 91-
thought you might like
a remembrance of your
Navy days.

Love you
Deb

DEDICATED
TO
THE MEN AND WOMEN
OF
THE PUGET SOUND NAVAL SHIPYARD

INTRODUCTION

NIPSIC TO NIMITZ is an exciting story of the United States Navy in the Pacific Northwest. From the pioneering days of the initial survey party and the establishment of the Puget Sound Naval Station in 1891, the hundred year history of what is now the Puget Sound Naval Shipyard is a tribute to the men and women, both civilian and military, who had the foresight and vision to plan for the future and the skill and determination to ensure that the Shipyard was fully cable of fulfilling its role in time of war and in insuring the peace.

The story begins with the singular determination of Lieutenant Ambrose Wyckoff, but its strength is the unfolding of the contribution of command leadership and the documentation of the accomplishments of each succeeding generation of the civilian workforce. It is the story of sails and coal to nuclear power, dry docks and shop facilities, ships overhauled and repaired. It is the history of the growth of the Shipyard's industrial strength in support of the needs of the United States Fleet.

Truly the Shipyard of today is far beyond what Lieutenant Wyckoff ever envisioned and provides a lesson for the next century, that the world of human accomplishment holds no bounds. The Shipyard is poised to continue improving its capabilities through innovation and hard work.

Arthur Clark
Captain, United States Navy

Captain Clark is Commander of the Puget Sound Naval Shipyard and has been selected for the rank of Rear Admiral.

NIPSIC to NIMITZ

A Centennial History of Puget Sound Naval Shipyard

By Louise M. Reh and Helen Lou Ross

NIPSIC TO NIMITZ

A Centennial History of Puget Sound Naval Shipyard

Library of Congress Catalog Card No. 91-70974

ISBN 0-931475-02-3

Publisher: Federal Managers' Association
Chapter 14, Bremerton, Washington

Art Work by Linda C. Deline

Printed in the United States of America

By: Bremerton Printing Company
Bremerton, Washington

Inside front cover U.S.S. NIPSIC
Larry Jacobson

Inside back cover U.S.S. NIMITZ
Steve Zugschwerdt, SUN

CONTENTS

FOREWORD

A naval shipyard without its people is simply a collection of buildings and equipment. With its people, a shipyard becomes alive and productive; it develops a personality and a reputation. The reader will find this history of Puget Sound Naval Shipyard not only traces in detail the changes on the waterfront as the shipyard grew, but also tells an interesting tale about the people and events that made it all happen.

Louise Reh and Helen Ross and their committee have done their research well. The reader will find that the Shipyard has had its high and low periods. There were times when international events forced recognition of the Shipyard as a vital asset. During these times capital investment in facilities and employment surged. Then came times when there appeared to be no external threat to our country and appreciation of the Shipyard waned. In the low periods, Shipyard leaders struggled to retain their skilled people.

For the naval history buff, NIPSIC TO NIMITZ provides informative and interesting reading. Sprinkled throughout are facts about evolutionary changes in naval ship systems design. Probably few know that the Shipyard was involved in advances such as the introduction of mechanical-electrical gun control systems during World War I and the development of steam catapults to accommodate heavier jet aircraft. By simply thumbing through the pages and looking at the incredible collection of photographs, one gets a quick overall appreciation for how rapidly ship design and technology have moved and how the Shipyard in this dynamic environment had to anticipate the facilities and training required to meet its mission.

It is, of course, impossible for a history like this to recognize each of the thousands of people who have contributed through the years to this great shipyard. However, the authors with painstaking effort have discovered names of many of the people involved, even identifying them in old photographs. In taking this walk back through time, there will be few families living in the surrounding communities, or in the Navy for that matter, who are not touched personally, as they recall events and individuals. The Shipyard has earned through its people a superb reputation and achieved a lasting bond with the Navy and its neighbors. This book tells it well. Enjoy.

Roger B. Horne Jr.
Rear Admiral United States Navy

Admiral Horne was Commander of the Puget Sound Naval Shipyard 1981-1984 and is presently Chief Engineer and Deputy Commander for Ship Design and Engineering, Naval Sea Systems Command, Washington, D.C.

PREFACE

The world has changed greatly since 1891 when Navy Lieutenant Ambrose Barkley Wyckoff established the Puget Sound Naval Station at Turner Point in Kitsap County, Washington. The steam engine, open hearth furnace and the rolling mill, amazing advancements of the 19th century, have given way to the sophisticated tools and microchips of the 20th century. The Puget Sound Naval Shipyard has kept pace, often leading the way with inventions and innovative applications which would astonish Wyckoff and his contemporaries.

The average citizen is not aware of the importance of the work done or the great contribution to the country this industrial giant's employees provide. People who have no direct contact with the Shipyard hear or read of a ship's arrival; they may see sturdy tug boats shepherd a ship through Rich Passage or into the Shipyard. Then a curtain drops and nothing is heard of the warmly welcomed ship until it departs.

This book reveals what goes on while a ship is at the Puget Sound Naval Shipyard. On these pages you will find the diversity of occupations and intensity of responsibilities in a naval shipyard. It cannot mention all the people or describe all the events that have given the Puget Sound Naval Shipyard its excellent reputation; it does give a look at many of them.

More than six years have been spent gathering information, writing and preparing the manuscript for publication. The most difficult task has been deciding what to use, what to omit. Eight-page manuscripts had to be condensed into a few paragraphs and some interviews are mentioned only with a line of type. Many stories are left for a later telling and the collection of documents will be saved for use by other researchers. The authors ask that anyone with questions or comments to make on the book write to them in care of the Kitsap County Historical Society, 3343 NW Byron Street, Silverdale, WA 98383

This book has been the work of many people. There are a few whose dedication merits special mention for their faithful and unwavering support: George Campbell, Dick and Doris Linkletter, Ralph Smith and Fred Timmerman. Others are thanked on page 263.

For all of us, the goal in writing this book has been to provide a lasting legacy of the Puget Sound Naval Shipyard's first one hundred years and to describe the ways in which PSNS has served the country in peace and war. Most importantly, this history has been written to be a source of pride and pleasure for all who have been associated with the Puget Sound Naval Shipyard.

The Centennial Book Committee

Louise M. Reh, Chairman
Helen Lou Ross
Pat Beatty Caldwell
James K. Herron
Frank J. Reh

Writing his reminiscences, Lieutenant Wyckoff stated, "On September 16, 1891, the residents of the neighborhood gathered at the point near the center of the present station, and all uncovered while I read my orders taking possession. When I had finished, my daugher Selah ran the Stars and Stripes up to the top of a tall tree which had been denuded of its branches.
(adapted from cover of September 14, 1951, SALUTE, by Linda Deline)

Chapter One

The Realization of a Dream

1877-1899

FRIDAY, APRIL 1, 1892 USS NIPSIC

Squally with occasional rain.

In obedience to orders from the Secretary of the Navy, Lt. A. B. Wyckoff, USN, assumed charge of the ship.

Light southerly airs all night, with overcast sky.

* * * * * *

Thus read the log of USS NIPSIC for April 1, 1892. With that brief notation, Navy Lieutenant Ambrose B. Wyckoff took command of the 30-year-old wooden gunboat that would serve him as office and home.[1]

Although NIPSIC had been battered unmercifully in a hurricane off Samoa in 1889 and later decommissioned, one would have thought her the finest ship afloat from the reception she received when she arrived in Puget Sound early in 1892. As NIPSIC and her escort, the gunboat USS MOHICAN, entered Elliott Bay, crowds swarmed Seattle's waterfront to view the ships and await the opportunity to go aboard. Small boats with flags flying escorted her across the Sound to a joyous reception in Sinclair Inlet.

On the Inlet's southern shore, the small town of Sidney welcomed the ship with a 21-gun salute given by an old cannon mounted on Fort Hill.[2] NIPSIC was regarded as assurance that the Navy was going to develop the Naval Station on the opposite shore of Sinclair Inlet.

Wyckoff had taken command of the Puget Sound Naval Station on September 16, 1891, as his daughter raised the flag over the first naval establishment in the Northwest. Wyckoff termed his new post "the consummation of my most earnest desire for 12 years past."[3] That station is now the Puget Sound Naval Shipyard.

In 1877, Wyckoff arrived at Seattle on temporary duty with a Coast and Geodetic Survey party. At that time Seattle had fewer than 3,000 inhabitants. Wyckoff later wrote:

> *Seattle was our headquarters for many months and we got well acquainted with all the old pioneers. They were a delightful lot of people and had both the time and inclination to enjoy themselves. The craze for the almighty dollar had not seized upon them and there were no cliques in society upon account of financial standing . . . Tacoma consisted of a few houses, and stumps were everywhere, even in some of the principal*

EARNEST, a U.S. Coast and Geodetic Survey vessel, became Wyckoff's headquarters in 1879 when he was placed in charge of the survey work on Puget Sound. Later EARNEST was transferred to the Navy and served as a lighter at the Puget Sound Naval Station until 1902.
Centennial Book Committee

streets . . . our winters were spent at Olympia where we computed and plotted the data obtained during the dry season. Olympia was at that time the largest city on the Sound. The society was made up of the refined and cultured families of the officers and ex-officers of Washington territory and it was very delightful.[4]

Wyckoff, impressed with the geographical position and natural resources of the region, became convinced that the Puget Sound area was the place for the principal naval establishment on the West Coast. Noting Puget Sound's great depth of water and lack of obstructions, its position in relation to Alaska and the Orient, and the temperate climate, Wyckoff declared, "Such a combination of advantages for a naval station could not be found elsewhere in the United States."[5]

Wyckoff recommended 200,000 acres of timber land be selected for a naval reservation. This was still the era of wooden ships, and he hoped the land would be set aside for Navy use as had the live oak forests of Florida. A bill was introduced in Congress in 1880, but nothing came of it. Wyckoff attributed this to the low ebb of naval affairs and general ignorance of the resources and advantages of the Puget Sound area.

Ordered to torpedo school at Newport, Rhode Island, in 1880, Wyckoff lobbied for the project. When ordered to China with the Asiatic Fleet, he continued to extol Puget Sound's merits as a Navy Yard site. So persistent were his appeals that he became known as "that Puget Sounder".

During the 1880s, Puget Sound became better known. The burgeoning lumber industry served as a catalyst for other commercial ventures, the Northern Pacific Railroad completed its lines to Tacoma, and the population increased considerably. In September 1888, John B. Allen, Congressional Representative for the Territory of Washington, introduced a bill in Congress authorizing President Grover Cleveland to appoint a commission to select a suitable site for a navy yard and dry dock in the Northwest.

Captain Alfred Thayer Mahan (whose book THE INFLUENCE OF SEAPOWER UPON HISTORY — 1660-1783 would be published in 1890) was the senior of three naval officers appointed to the commission. They were ordered to examine the Pacific Coast above the 42nd parallel in the State of Oregon and Territories of Washington and Alaska.

The Commission decided that Puget Sound was the most suitable location for a navy yard and dry dock in the entire region. It recommended the immediate purchase of 1,752.2 acres of land on Point Turner between Sinclair Inlet and Port Washington Narrows.

The report engendered extreme opposition in the East. In addition, Oregon and California would not support the purchase. Oregon hoped to have a yard and dock built there, and California feared future competition with its Navy Yard at Mare Island.

Congress would not approve the purchase. However,

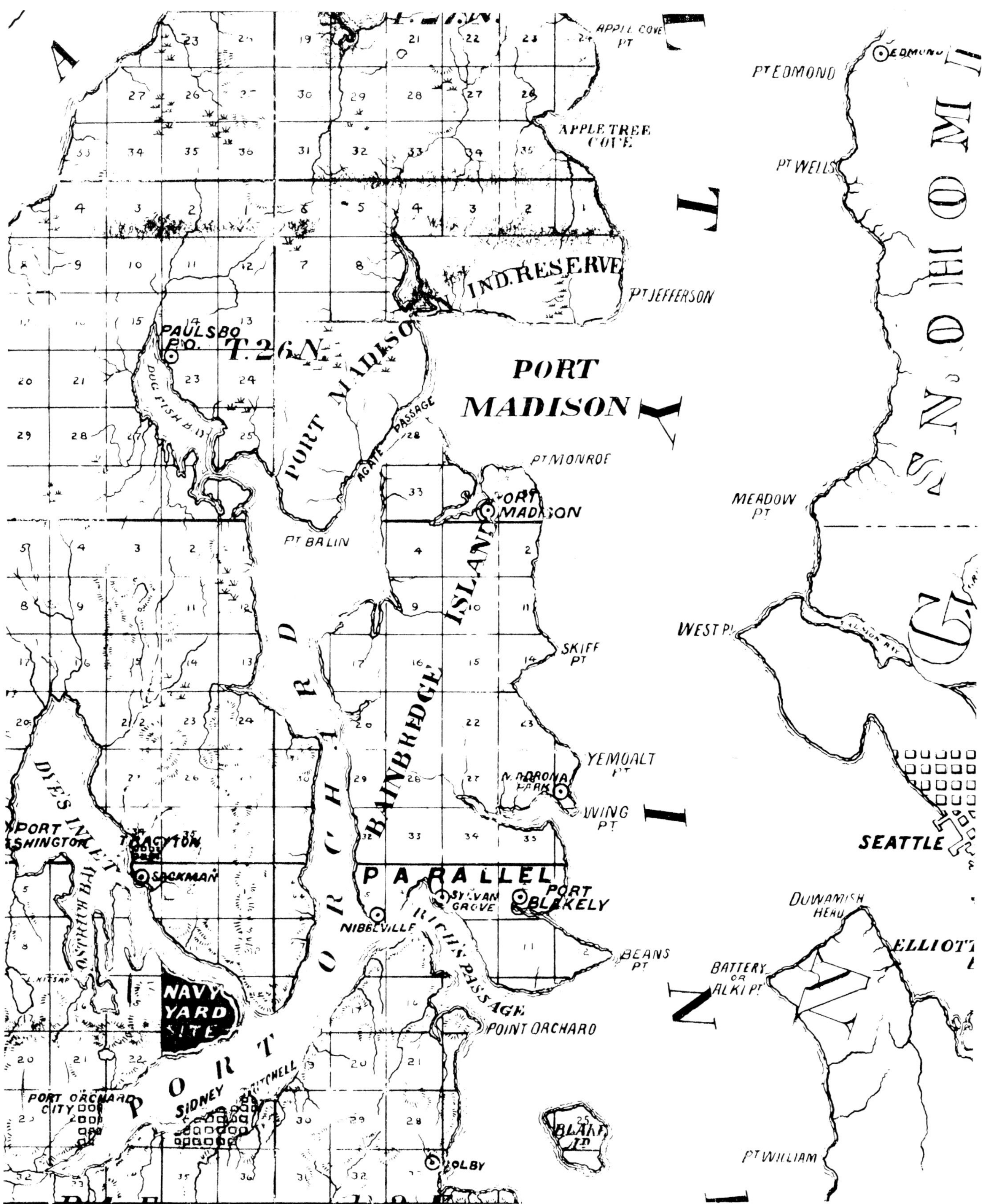

Section of map of Kitsap County, circa 1890, labeled "compiled from official records by O. P. Anderson Co. (Inc.) Engineers and Draftsmen, Butler Block, Seattle." The defensibility of the Navy Yard site from seaborne attack is graphically shown in this map which also shows the relationship of the site to the City of Seattle. Note location of Sidney (now Port Orchard), Port Orchard (later Charleston and now Bremerton), Tracyton and Nibbeville.

Centennial Book Committee

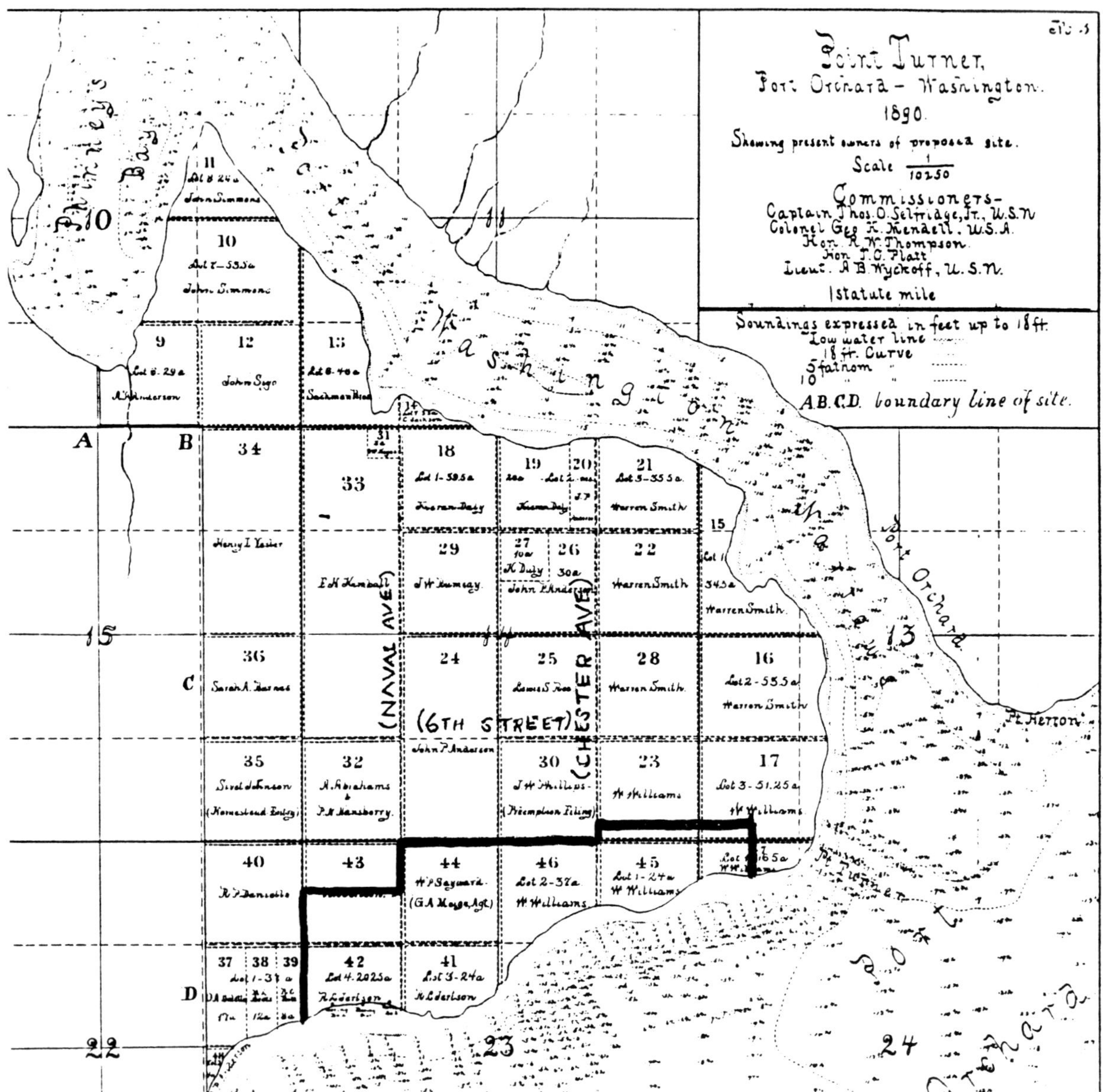

The 1890 government commission recommended acquisition of 1,716 acres on Point Turner from Phinney Bay on the North to Sinclair Inlet on the South, plus land around Kitsap Lake. The map is marked to show the area purchased in 1891 and 1892. Kitsap County Historical Society

when Washington became a state in November 1889, Allen, now a senator, was able to add to the Naval Appropriations Bill money for a commission to select a site, this time solely for a dry dock. This second commission consisted of two civilians, an army officer, and two naval officers, one of whom was Wyckoff.

This group chose practically the same area as the Mahan Commission and recommended the purchase of 1,716 acres. The value of the land was set at $36,638. Their report, dated December 23, 1890, failed to move Congress. The Navy Department was beginning to build modern ships, but in the interests of economy Congress had closed all Navy Yards except those at Brooklyn, Norfolk, Mare Island and Washington, D.C. Building a dry dock in a little understood region was considered extreme folly.

Senator Allen then proposed an amendment to the Appropriations Act to include funds for a dry dock for commercial as well as naval purposes. In debate, much was made of the fact that the United States had no drydock north of San Francisco large enough to accommodate the country's larger commercial sailing vessels and steamships. Often ships in need of repairs had to be sent to the British Columbia Dock Yard at Esquimalt.

Recognizing the resulting flight of American money to a foreign port, Congress took another look at the proposal. Senator Allen was able to line up sufficient votes for passage in the Senate. Passage in the House was another matter. Hilary A. Herbert (Democrat-Alabama) of the House naval committee, bitterly opposed the amendment. However, according to Wyckoff:

> *During the confusion on the floor incident to the closing days of Congress, it was put to vote and carried before he* (Herbert) *could demand the ayes and noes, which would probably have defeated it, so that the victory for our dry dock was won by the skin of our teeth.*[6]

Ambrose B. Wyckoff, born in 1848 near Jersey City, Illinois, entered the Naval Academy in 1864, when it was located at Newport, Rhode Island. In addition to founding the Puget Sound Naval Shipyard, he is noted for charting the waters of Elliott and Commencement Bays and for directing the Coast and Geodetic Survey's Hydrographic Office in Port Townsend. Legislation introduced at his instigation provided the first funds for the Lake Washington Ship Canal. Wyckoff died in Ontario, California, on Memorial Day in 1922. During Washington's Centennial Celebration in 1989, the Washington State Historical Society placed Wyckoff's name in its Hall of Honor. Kitsap County Historical Society

Wyckoff, assigned special duty on the staff of Secretary of the Navy Benjamin F. Tracy, was ordered to proceed immediately to Puget Sound and select the site for a drydock for naval and commercial purposes. The maximum amount of land allowed was 200 acres, to be purchased for no more than $10,000. The dry dock had to be no less than 600 feet long and 70 feet wide and admit vessels drawing 30 feet of water. It was to cost no more than $700,000.

Wyckoff arrived in Seattle on April 2, 1891. One can imagine the enthusiasm he felt and exhibited as his long cherished dream neared realization. "That Puget Sounder" had returned, this time to talk business.

Two days later Wyckoff met with the leaders of the little town of Sidney on the south shore of Sinclair Inlet. He immediately made his plans clear: he would pay only $50 an acre since he was determined to have the maximum amount of land authorized. He specified his interest in the land on the north side of Sinclair Inlet. Two committees were formed, one to meet with the resident owners of that property, the other to meet with William Bremer of Seattle, a realtor who had financial interests in the area.[7]

Because of the limited appropriation provided, the land selected was only a small portion of that recommended by the commissions. Stretched along more than one mile of tidelands, it was a wilderness except for the small Robert H. Jertson family home. Broad marshes at both ends of the tract extended far inland.

> It was *covered by a dense tangle of swamp grass, alders, giant cedars, and marshwillows. Between these marshes were two ridges which jutted into the bay. The ridges were covered by a heavy growth of huge firs. Madronas grew quite plentifully. Rhododendron, dogwood, wild currant and many other flowering bushes and trees grew over the hillside.*[8]

The interest shown in the area by the government commissions caused a boom, and land sold for over $100 an acre. When landowners refused to sell at the lower price, Wyckoff very openly looked at a site on Dog Fish Bay. Pioneers' memoirs state that decisions to sell to the government for $50 an acre were influenced by gifts of money or land from Sidney residents, who were determined that the dry dock would be built on Sinclair Inlet.[9]

The property Wyckoff bought was owned by a number of individuals. Contrary to popular belief, William Bremer did not own any of this land at that time. The February 1891 title for the easternmost acreage on Point Turner shows its owner to be Henry Hensel, later Bremer's brother-in-law. Wyckoff arranged for the purchase of 81 acres from Hensel, 42.87 acres from the Robert Jertsons and 21.38 from the Beruth Olsens, who recently had purchased it from the Jertsons.

These 145 acres were the lands acquired when Wyckoff took command of the Puget Sound Naval Station. The 40 acres of high ground known as the Sayward Tract were finally obtained through condemnation proceedings made necessary by question of clear title.

Wyckoff received an extension of his orders that allowed him to supervise the start of work on the dry dock. When test holes indicated quicksand in the Jertson (western) basin, it was decided the dock should be built in the eastern basin. This necessitated the purchase of another five acres, obtained from William Bremer in March 1892. Bremer had purchased all the Hensel land outside the station and platted the town of Bremerton in December 1891.

The political subterfuge of a single drydock served as the nucleus for a station that would grow to be one of the nation's foremost shipyards. Selah Wyckoff's act of running the Stars and Stripes to the top of a de-branched fir tree on September 16, 1891, he signalled the beginning of the great enterprise.

Wyckoff had done well. He had secured 190.25 acres of land for $9,512.50.

> *My entire expenses, for travel, legal fees, abstracts, recording deeds, etc., was seventy-four dollars and seventy-five cents in securing the cheapest and best site for a navy yard the government will ever possess.*[10]

The Lieutenant's work had just begun. December 1891 orders from the Chief of the Bureau of Yards and Docks requested a full and complete plan for development of the Station. Envisioning a first class construction and repair yard, the Secretary demanded a thorough study.

Because of the shortage of housing in the area, Wyckoff set up his office and quarters on the old Coast Survey ship YUKON. He found her in Eagle Harbor, Bainbridge Island, with several feet of greasy water in her bilge. Although he hired four men to clean her out, it was several days before anyone could sleep below because of the foul bilge water.

Living conditions on YUKON were never satisfactory and Wyckoff eagerly awaited NIPSIC's arrival. With NIPSIC finally anchored in Sinclair Inlet and outfitted for use as office and quarters, Wyckoff and his three daughters moved aboard.

The east end of Puget Sound Naval Station on Turner Point in 1892. Two poles at left of picture indicate where dry dock will be built. Kitsap County Historical Society

Civil War survivor, USS NIPSIC, laid down in 1862, contributed to the Confederacy's defeat by participating in the blockade of the South's ports. During her long life, she sailed with the South Atlantic Squadron off the coast of Brazil, in the Mediterranean, along the coast of Africa and as the Station Ship in Apia Harbor, Samoa. As a gunboat, her principal task was to guard American commerce and interests.

Photo taken by Wyckoff family from deck of USS Yukon. Kitsap County Historical Society

Stella and Selah Wyckoff aboard USS NIPSIC. Because of their mother's ill health, Stella, Selah and their younger sister, Carrie, were raised by family members. The few months on NIPSIC was the only time Wyckoff had all three girls with him at the same time. Only Stella remained in Washington. She taught German at the University of Washington, married and made her home in Seattle.

Centennial Book Committee

Wyckoff's staff was small: Civil Engineer Thomas McCallum USN, Asst. Paymaster E. B. Webster USN, Chief Clerk William E. Wilmerding, Draftsman F. W. D. Holbrook,[11] Axeman Daniel Lund[12] and Wyckoff's boyhood friend, George Terrell.[13] For over a year Terrell kept track of the daily flood and ebb tides with the use of a gauge fastened to a piling on the wharf. From this information, Wyckoff decided on the depth of the dry dock.

Construction of the dry dock was of paramount importance. Advertisements for bids were sent to contractors on both coasts. The contract was awarded to the Tacoma firm of Byron Barlow and Company on October 29, 1892, for a bid of $491,465.

Ground was broken on December 10. The official party consisted of Wyckoff, Civil Engineer U.S.G. White, Paymaster Webster and contractor Byron Barlow. The ceremony, attended by some 100 people from surrounding towns, opened with Wyckoff narrating the history of the enterprise. He spoke glowingly of the wonderful site, declaring it the best possible location for a great naval station.

His second daughter had raised the 44-star flag signalling the establishment of the Station. Now, his eldest daughter, Stella, turned the first spadeful of dirt for the symbolic start of the dry dock. Wyckoff turned over temporary possession of the dry dock site to the contractor.

Now that the dry dock was on its way to reality, Wyckoff looked to his health. Though racked with inflammatory rheumatism, which he attributed to living on YUKON, he had completed his task. He wrote in his journal:

> *The honor and responsibility of the assignment to a junior officer was unprecedented in the Navy and my professional pride made me remain until the duty was fully accomplished, not withstanding my painful illness.*[14]

Lieutenant Commander James C. Morong relieved Wyckoff of command on February 6, 1893. Wyckoff left for the Army and Navy hospital in Hot Springs, Arkansas. He was due for promotion but the Medical Examining Board ruled he would never again be physically fit for sea duty.

Benjamin Tracy wrote a letter on Wyckoff's behalf to his successor as Secretary of the Navy, Hilary A. Herbert, but to no avail. On July 3, 1893, Wyckoff was placed on the retired list, much against his wishes.

One of Morong's first acts was to make NIPSIC more accessible to the Station by moving her closer in and connecting her to the shore with a floating pier. He also ordered construction of a roof from the poop to the forecastle and considerable repairs and painting.

Civil Engineer White and his crew worked in the former Jertson home, a mile west of the dry dock. In 1894 they moved into newly completed Building 51, located just west of the dry dock.

Before buildings were erected on the Station, NIPSIC not only served as quarters for Wyckoff's family, but also as offices for the Station. When Lieutenant Commander Morong commanded the Station, NIPSIC was moved and moored at the end of a pier.
Centennial Book Committee

The Civil Engineer set up his office in the former Jertson home. Seated beside the building are: (l to r) A. F. Haynes, T. F. Pearl (holding "Gyp"), Civil Engineer U. S. G. White, USN, and David Clopton. Standing are F. W. D. Holbrook, "Pete" W. A. Smith, and George M. Terrell. To the far right is dog called "Penwiper." Note the state-of-the-art surveying instruments.

Kitsap County Historical Society

August 1, 1894. Foot by foot, the Station's first dry dock takes shape. An enormous number of pilings are swung into place; the winch is powered by a donkey boiler stationed under cover in mid-dock. In a letter of January 21, 1893, Wyckoff gave the contractor authorization to cut firewood on the station "provided you agree that all refuse matter be burned, the wood used entirely for purposes in connection with construction of the dry dock and without cost to the Government".

Puget Sound Naval Shipyard

The basic steel structure of the caisson (gate) for Dry Dock 1 was manufactured in Newburgh, New York. A March 1894 communication from Civil Engineer P. C. Asserson at the New York Navy Yard to the Chief of the Bureau of Yards and Docks in Washington D.C. reads: "A dry dock caisson constructed . . . for the dry dock at Puget Sound has been completed as far as is practicable before its final completion at the dock . . . The caisson with all appurtenances has been shipped on February 28, 1894, in the ship HENRY B. HYDE."

Puget Sound Naval Shipyard

Moran Brothers, who later built the battleship USS NEBRASKA, constructed Building 52, the pump house for Dry Dock 1. This 1895 picture shows the dock's wooden sides and the granite sill. The man standing in the foreground gives an idea of the size of the granite blocks.

Centennial Book Committee

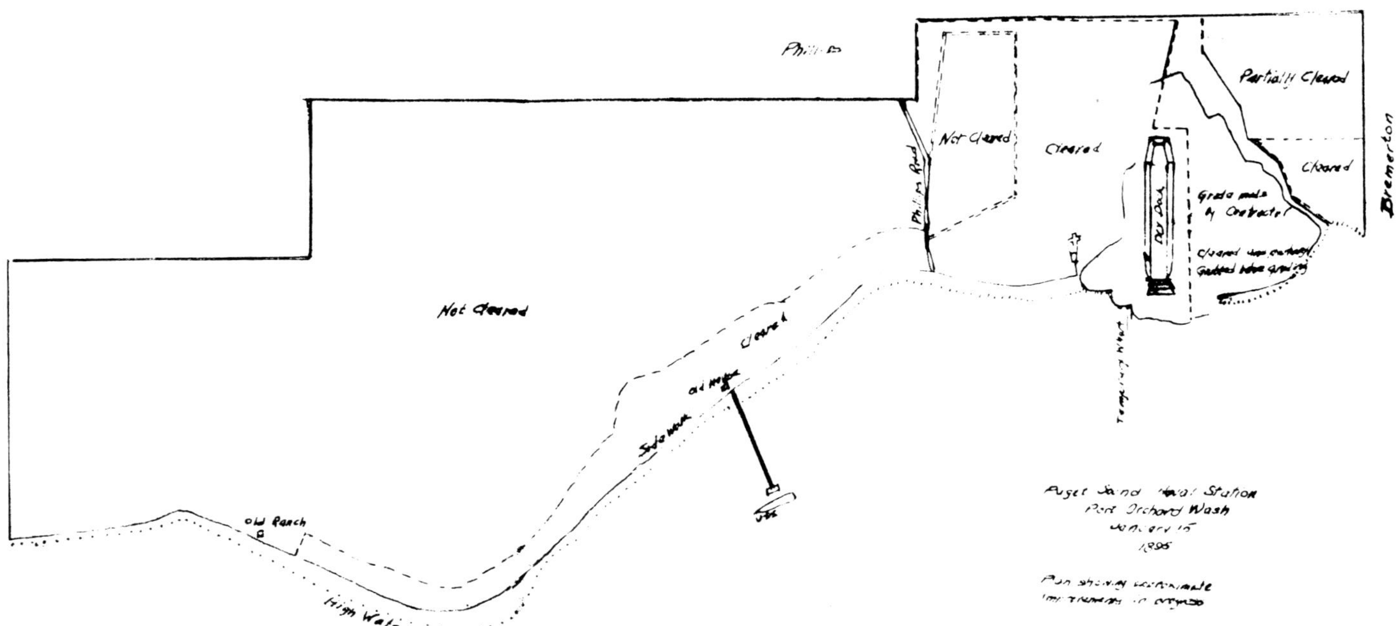

Only a small portion of land had been cleared when Civil Engineer Hollyday drew this map in 1895. He recorded Dry Dock 1 as under construction, and the wooden sidewalk (built by dock contractor) paralleling the waterfront from Charleston to the dry dock site. "USS" at dock's end refers to USS NIPSIC. Puget Sound Naval Shipyard

The Station's first dry dock was constructed with wooden sides and keel blocks, except for the outboard 70 feet which was made of granite blocks. To the left of the dock stands the first power plant where Engineer James Gibboney was in charge. Dry Dock 1 was rebuilt in 1931, concrete replacing wood. Puget Sound Naval Shipyard

Robert Moran and Company of Seattle built the dry dock's pumping plant. The plant boasted three large centrifugal pumps which could lift more than 110,000 gallons of water per minute.

When it was completed in April 1896, the dry dock was 650 feet in length, 130 feet in width and 39 feet in depth, making it the largest U.S. naval dry dock.[15] Because of the destructive action of the teredo worm on wood, the first 70 feet of the entrance to the wooden dry dock were built of stone.

When the dock was ready for testing on April 23, steamers brought crowds of sightseers from Seattle and Tacoma. At 9 a.m. the coast defense monitor MONTEREY cast loose from her buoy in Sinclair Inlet and entered the mouth of the dry dock, across which a blue ribbon stretched. A Seattle newspaper reported:

The great ship daintily put her nose against the silken barrier, as gracefully as a blooded racer, gently but panting to try its mettle; the ribbon parted amidst a murmur, a passing ripple . . . and the monitor glided into the dock."[16]

Naval officials, technical experts and civilians watched from along the coping. With the caisson in place, the pumps emptied the dry dock of its 14,000,000 gallons of water. Wyckoff had returned for the occasion. He, Morong and ex-Senator Allen rejoiced as the big ship settled on the long row of keel blocks, the hull became visible and the floor of the dock emerged.

A telegram sent to the Bureau of Yards and Docks read, "MONTEREY docked today with complete success." The terseness of the message concealed the delight of all concerned. That day, everything in the Station, including machinery, was open to inspection by the public. The boiler-installing company hosted the visitors at lunch on the steamer MARY F. PERLEY.

Navy coast defense monitor, USS MONTEREY, breaks the blue ribbon to be the first ship to enter the new dry dock. At the time, Dry Dock 1 was probably the third largest in the world. *Kitsap County Historical Society*

Erecting the Construction and Repair Shop Building (Bldg. 58) took nearly a full year. Engineer James Gibboney's log reports the arrival of trusses in September 1896, installation of machinery in April 1897 and the assignment of the first machinist, Thomas Byrne, 11 days later. Testing and operation of the machinery began in July 1897.

Kitsap County Historical Society

Thorough testing was carried out, and on September 29 the Navy accepted the dock. The following day James Gibboney was placed in charge of the pumping plant. Gibboney had been hired in 1894 as a sub-inspector during the construction period and proved so competent he was promoted to engineer.

Visitors, dignitaries and excursionists alike, continued coming to view the new dry dock. Steamers unloaded large groups; one day the members of the State Bar Association, another day 150 excursionists from a lumbermen's group.

In 1894 Morong hired Frederick G. Forbes as a special laborer. Forbes' work for Byron Barlow had impressed Morong and soon he was appointed Chief Clerk to the Commandant, a position he held until his retirement in the early 1930s.[17]

Morong's staff grew with the addition of another officer in 1895, when Passed-Assistant Surgeon James Stoughton set up sick quarters on NIPSIC. Morong had pleaded regularly for a doctor for the Station as the nearest doctor was in Sidney and his territory covered a large portion of Kitsap Peninsula.

In addition to the dry dock, Morong's responsibility included development of the rest of the Station. Morong wrote regularly to the Bureau of the need for a proper water supply system. Samples of water from various sites were sent to Washington, D.C., for testing. In August 1895 the firm of Joslyn and Gibson of Seattle was hired to dig the well.

That same month the Seattle architectural firm of Chamberlain and Siebrand received the contract to design an office building and five officers' quarters. The site for the office building (Bldg. 50) was on a mound 35 feet above high tide and 500 feet west of the dry dock. The five homes were located on a hill about 1,800 feet west of the dry dock and 75 feet above high tide. Zinndorf Construction Company of Seattle received the contract to build the office and quarters. The six buildings were completed in 1896 at a total cost of $33,000.

1897 photograph of first five Officers' Quarters. Built of lumber from Port Blakely mill, these quarters are 1800 feet west of Dry Dock 1. The office building of that time (Bldg. 50), was located on a 35-foot rise 500 feet west of the dock. Within a decade, landscaping and natural growth of vegetation softened the residential area's raw appearance.

University of Washington

The Marine Barracks and Officers' Quarters were built in 1899 near what is now the corner of Gregory Way and Chester Avenue. The Barracks was razed during the 1919-1920 regrade and the quarters moved to its present location near the State Avenue gate.

Kitsap County Historical Society

USS OREGON, one of the new battleships, was the pride of the fleet at the turn of the century. The ship had just completed a six weeks availability for the installation of new bilge keels to improve her stability. When the USS MAINE blew up in Havana Harbor, OREGON was in Dry Dock 1 at the Puget Sound Naval Station. She quickly undocked and sailed to San Francisco. From San Francisco she raced nearly 15,000 miles in 66 days around Cape Horn to Key West, Florida, where she joined the North Atlantic Squadron en route to Cuba. Puget Sound Naval Shipyard

Captain William H. Whiting became the next Commandant. The Whiting family, the first to live in Quarters C, occupied the largest of the five homes.[18] Their spacious home had the potential for gracious living, but the surroundings were primitive. On August 8, 1896, the day the Whitings arrived at the Station, a huge fire threatened the destruction of most of the buildings. The next morning, Mrs. Whiting roused the family with her screams. There was a bear outside the back door.[19]

Marines, with a sergeant in charge, had provided security on the station for several years. During Captain James G. Green's 1897-1899 tour as commandant, the Marine Corps built an officers' quarters and barracks on the hill north of the dry dock. First Lieutenant E. K. Cole was in charge when the buildings were completed in 1899.

When the battleship USS MAINE blew up in Havana Harbor on February 15, 1898, the battleship USS OREGON was in the station's dry dock. Commissioned in 1896, OREGON was sent to Puget Sound for work because the dry dock at Mare Island Navy Yard was not large enough to hold her. OREGON entered the dry dock for installation of bilge keels on January 4, 1898; eight days later, 3,000 holes had been drilled and tapped into her hull for the bilge keels.

The last bilge keel plate was drawn up and bolted fast for riveting on January 28. Stormy weather prevented her scheduled undocking on February 15, MAINE's disaster date, but the next day OREGON was in the water. She took on coal and headed for San Francisco. From there she began her epic voyage around South America to join the war in Cuba. The Station's morale was high, knowing it had conditioned the ship to greatly increase her speed and stability.

After the Spanish-American War, the Station's future looked bleak. There had been only 12 dockings during the initial 33 months of the dry dock's existence, but with the exception of work done on the battleship OREGON, the availabilities were for less than two weeks. The dry dock stood empty for months at a time.[20] Men were hired to work when a ship came to the Station and then laid off until the next ship arrived. Skilled workers sent up from Mare Island to assist had little incentive to stay.

Fair | 74° | Monday July 10th 1899.
No Pumping | Water in dock stands at 12'4" mark
Hauled 12 large loads Coal

Tuesday July 11th
"Iowa" quit using our steam at 12:50 P.M. and disconnected pipe at 4 P.M.
Flooding dock at 5:32 P.M. Tide 26' Began to Pump at 6:25 P.M. Tide 27'10" and lower compartment full.
"Iowa" undocked at 8:10 P.M. amid the shouts of 2500 people and toots of whistles from the "Skagit Chief", "State of Washington" and "Alice Gertrude" that were floating beyond the buoys.
Berglind off at 8:30 P.M. Harrigan worked 4:30 to 10 P.M.
Reynolds & Knutsen off at 8:05 P.M.

Warm | 80° | Wednesday July 12th.
Harrigan out at 4 A.M. to get up steam by 7 a.m.
Case, Berglind & Knutsen & I out at 6 a.m. Reynolds out at 7 a.m.
Harrigan in Chg. of Boilers.
Caisson sealed at 7:35 A.M. and 3 Main Pumps going at 7:45 A.M.
Dock dry at 9:45 A.M. Cut out all boilers but 1 & 2.

Warm | 79° | Thursday July 13th
Put Harrigan in Chg. of boilers.
D.P. run 7 to 9 A.M.c & 3 to 4 P.M.k Berglind & Reynolds cleaning engines.
Anderson helping Harrigan at boilers.

Warm | 81° | Friday July 14th
D.P. run 7 to 9 A.M.c & 3 to 4 P.M.k
Knutsen helping Berglind to take of Crosshead of Eng. #3 which became loose during last run.

80° | Saturday. July 15th.
D.P. Run 7 to 8:30 A.M.c & 3 to 4:45 P.M.k Hauled 4 loads Coal weighing 7000 lbs. each according to Holt's weight the bed is rounded up!
Berglind & Knutsen making drift key to back off Crosshead #3.

Very warm | 80° | Sunday, July 16th.
Nobody working.

A page taken from the 1899 Pumping Plant Log, as recorded by Engineer Gibboney, reveals some types of work performed at the plant and dry dock. Besides noting dockings and undockings, coal hauling, workers' hours and 80 degree days, the record shows "Nobody working" on a "very warm 80° Sunday, July 16th." Puget Sound Naval Shipyard

Green was completing his tour when word came that the new Chief of the Bureau of Yards and Docks, Rear Admiral Mordecai Endicott, had convinced Secretary of the Navy John D. Long that Puget Sound was not a good place for a naval station and it should not be developed further. The main objection to the area was the shortage of trained workers. That situation was exacerbated by the discovery of gold in the Yukon and Klondike, which caused a general exodus of men from the Pacific Northwest.

At this critical period, Wyckoff returned to the area and was dismayed by the indifference of the people and the doleful atmosphere. He appeared before the Seattle Chamber of Commerce to rally support. At Wyckoff's instigation, the Chamber appointed a committee, chaired by Allen and including Judge Thomas Burke and Congressman E. O. Graves to investigate the situation.

According to Wyckoff, their report of March 28, 1899,

> *. . . bristled with facts so cogent and lucid all further doubt as to the eminent fitness of the location was removed. On the contrary, the committee so ably set forth the wonderful natural advantages of the present site and compared it so favorably with all existing Navy Yards and Stations that a veritable wave of enthusiasm was created among both press and people of the sound . . .*[21]

Washington's Congressional members procured appropriations of $300,000 for work at the Station in 1900 and almost $500,000 in 1901. During the last nine years of the nineteenth century, the waterfront at Point Turner was transformed from a forest into a naval station with great potential.

It was only the beginning.

The Seattle Post Intelligencer on April 9, 1899, used this picture in an article urging further development of the Puget Sound Naval Station. Of the advances already made, the boardwalk through the Station and NIPSIC anchored off shore both show clearly, while the General Office Building, Dry Dock Pumping Plant and the Station's wharf are recognizable in the background.

Kitsap County Historical Society

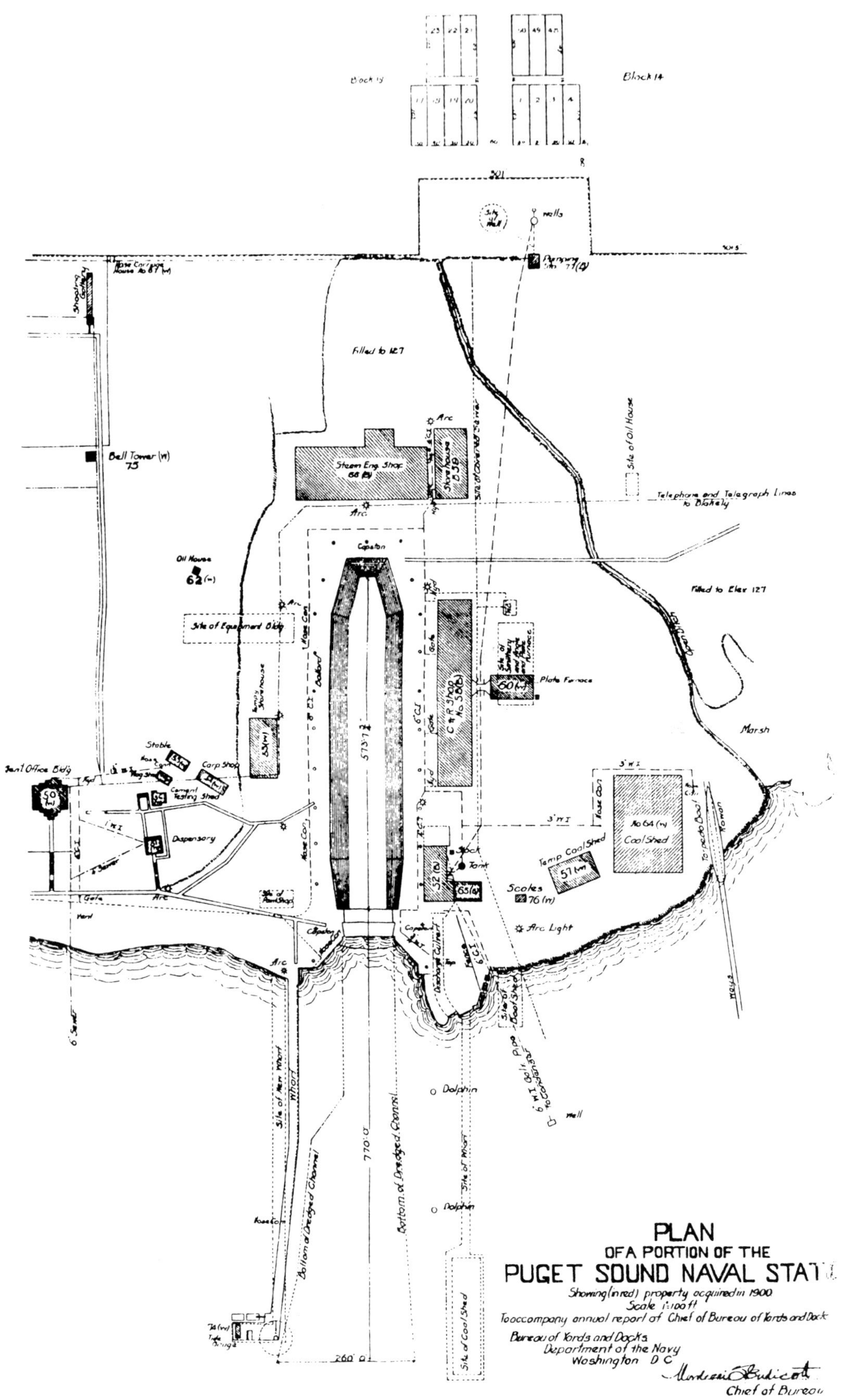

When Commander William T. Burwell took command of the Puget Sound Naval Station in 1900, the Yard had developed according to this map, which shows several features not in existence today. These include (clockwise from upper left) a Marine shooting gallery, long ditch through the marsh, telephone and telegraph lines to Port Blakely, torpedo-boat ways and arc lights on the waterfront. Letters following building numbers indicate whether built of B(rick) or W(ood).

Centennial Book Committee

Chapter Two

A New Beginning

1900-1915

Prosperity predicted by Wyckoff for the Sinclair Inlet towns arrived after the turn of the century. Towns at either end of the Station had struggled for existence for more than a decade. At the east end lay Bremerton, platted in December 1891. At the west end, two 1891 plats had been consolidated as the town of Port Orchard and incorporated in 1893.[1]

Population outside the Station grew sporadically; the 1900 Federal Census showed only 1,000 residents in the Sinclair and Dyes Inlet area. However by 1901, Bremerton alone could claim 500 citizens, enough to apply for incorporation.

The July 1900 arrival of a dynamic commanding officer, Commander William T. Burwell, brought new promise to the Station and surrounding communities. This promise began to take shape on July 23, 1901, when Navy General Order No. 56 changed the name of the docking station to Navy Yard, Puget Sound.

Burwell and later Commandants insisted on this official arrangement of the name, but Puget Sound Navy Yard was the popular name and is so used in this volume.[2] Whichever name was used, the people were proud of the Station's new rank.

One of the first improvements during Burwell's tour was the erection of a high wooden fence to replace the four strands of wire stretched from tree to tree to define the Station's perimeter. Marines at gates at the east and west ends further limited easy entry, but townspeople were allowed the convenience of using the boardwalk along the Yard's waterfront as passage between Bremerton and Charleston.

Small boys, avoiding a long walk to either gate, managed to find convenient ways to enter what was for them a fascinating playground. Ray Raines who grew up in Bremerton, described it thus.[3]

. . . This fence was too high for small boys to scale. However, like rodents, we found locations where we could tunnel under. We selected sites in brushy areas which were obscured from the view of Marine sentries who occasionally patrolled the fence and would have reported such breaches so they could be blocked with boulders. The tunnels were especially convenient, and time saving, when we wanted to go to the Receiving Ship pier. This pier was a considerable distance from the shops and dry docks . . . It was in a quiet area and the dock extended out from a sandy beach — a real nice place for boys to "fool around".

Acting Warrant Machinist Jesse Ely Jones, head of the Steam Engineering Department in 1900, and his wife proudly display son, Jesse Eugene, as they stroll along the Station's boardwalk. Kitsap County Historical Society

NIPSIC, serving as the Yard's Receiving Ship, was moored at this pier to the west of the industrial area. It housed recruits and sailors awaiting transfer to other duties. In 1900 NIPSIC also served as quarters for Acting Machinist Jesse Ely Jones, the first representative of the Steam Engineering Department in the Yard.[4]

With the increase in work, more workmen established homes in the area. The Wall family lived in the second house built in Manette, the quiet community across Washington Narrows from Bremerton. August Wall was the first coppersmith at the Naval Station in 1900 and his 16-year-old son, Walter, joined him in the shop the following year.

Walter Wall described their early day commute to work thus:

> *We'd grab a bite of breakfast, then the lunch pails, run for the beach, drag the boat to the water and get in. Out would come the oars, then row for all we were worth. Finally we would land and drag the boat up and tie it to a tree, hide the oars, run for the Navy Yard, drop our muster checks and start the day's work.*[5]

Painter Thomas Bright,[6] settled in Bremerton in 1900 on his arrival from California to work at the Station. His son, Thomas, at age 14, became the first apprentice molder for the Steam Engineering Department.

Jim Nelson started the Construction and Repair Department's Foundry in 1900. Fred Dahlquist was the department's first apprentice molder and later the second supervisor of the shop.

The changes in the appearance of the Yard reflected its higher status. The Construction and Repair Department's shops, (Bldg. 58),[7] stretched along the east side of the dry dock with the Department of Steam Engineering's shops

This view of the Naval Station was taken from the west on June 3, 1900. The first Machine Shop (Bldg. 66) is under construction in the foreground, with the storehouse (Bldg. 59) behind it. The dry dock and Construction and Repair Shops (Bldg. 58) are on the edge of the picture to the right. Scattered structures in background are in the town of Bremerton. Photo was made from a glass negative.

Kitsap County Historical Society

(Bldg. 66) and a brick storehouse (Bldg. 59) to the north. The Equipment Department's building (Bldg. 78) was under construction on the dry dock's west side.

Although buildings in the Yard were wired for electricity when built,[8] oil lamps provided the shop lighting for years. The General Electric Company constructed a 36 by 52-foot brick addition to the pumping plant in October 1898 to provide electricity. The system spread as the Yard received more appropriations.

The BREMERTON NEWS reported on October 19, 1901, that there were four arc lights on the wharf and one incandescent light in the storehouse. Four months later the SEATTLE TIMES reported the Yard's system had a 150-horsepower compound engine providing power to run two dynamos of 110 volts each to furnish electrical power for a three-wire system of 81 arc lights of 2,000 candlepower each, and 1,057 incandescent lights of 16 candlepower each.

The SEATTLE TIMES reported on February 9, 1902, that the Yard's:

> *. . . fresh water is obtained from a well fifty feet in diameter and twenty-two feet deep, which provides an inexhaustible supply of excellent water . . . upon the highest point of the grounds are located two large steel tanks, each thirty feet in diameter and forty feet in height with a capacity of 400,000 gallons . . . There are ten hydrants in the fresh water system.*

There were also 23 hydrants in a salt water system built for fire protection in the Navy Yard; one hydrant at the Bremerton gate was available for use by the town. Fire was a constant threat.

All the Yard's departments were busy. During the first two years of the twentieth century, 22 ships occupied the

Two years made a great difference. This 1902 photo, part of a panorama taken in 1902 from the mainmast of USS OREGON by Asahel Curtis, shows completed Building 66, as well as the increased size of the town. Centennial Book Committee

drydock. Battleships OREGON, WISCONSIN and IOWA took their turns with torpedo boats, civilian steamers and tugs and Coast Guard ships.

USS OREGON arrived on July 6, 1901, for repair of damages incurred when she ran aground in a fog in Chinese waters. The repair work at the Navy Yard consisted of removal and straightening or replacement of a large amount of bottom plating, plus installation of new weather decks.

Praise for Burwell filled the newspapers in 1902 as his departure drew near. James B. Miekle of the Seattle Chamber of Commerce wrote a lengthy article in the February 9, 1902, Seattle Times, in which he praised Burwell's energy, perseverence and accomplishments during his two years in command. Miekle described the Yard:

> *There is a complete marine railway adjoining the Construction and Repair shop for handling small-sized vessels. A Yard railway is now being built, running from the new wharf to the east of the Construction and Repair shop, through the Yard, connecting with all the principal buildings, and thence along the west side of the dry dock to the outer end of the freight wharf. This railway will add very greatly to the facilities of the Yard in the handling of heavy freight. The grounds have been improved and beautified by clearing the land, building roads and walks and transforming the rough land, which was covered with stumps and brush, into beautiful lawns and gardens.*

Burwell left the Navy Yard in August 1902 to become Commanding Officer of USS OREGON. His successor was Rear Admiral Yates Stirling Sr. In June 1903, Captain Charles J. Barclay became the Yard's eighth commandant.

Construction work continued. In 1903 the Equipment Department took over Building 78 on the west side of the dry dock. An addition to the General Office Building almost

The western part of Asahel Curtis' panorama shows additional progress. Sections of the Yard's fence are visible in the right background. Partially hidden by smoke are the officers' quarters. Buildings on the left waterfront were used in connection with the Receiving Ship pier. Land fill is visible to the right. Centennial Book Committee

doubled that structure's size. Two brick ammunition buildings and a home, Quarters J,[9] for the Warrant Gunner were completed near the mouth of the ravine in the center of the Yard. A coaling station, with a wharf capable of berthing and coaling two battleships at the same time, operated at the west end of the Yard.

The Yard's first hospital opened early in 1903, after the yard force added a two-story structure to the north side of a single-story wooden building put up by a contractor the year before. One surgeon and two hospital stewards attended the patients, whose numbers quickly outgrew the 16-bed facility. The overflow was housed in wood-floored tents.

A large number of patients with scarlet fever, diphtheria and smallpox added to the hospital's housing problems. In addition to these and other contagious diseases, the staff contended with problems related to the abuse of alcohol.

The prohibitionist movement was strong throughout the country, and the January 1903 SEATTLE TIMES headlined: "No more war vessels will berth at the Navy Yard until all saloons and disorderly houses are closed." The Navy Department objected strongly to the number of saloons in the Yard's vicinity and threatened to withhold all work. Bremerton complained the situation was not nearly as serious as that at Vallejo, adjoining the Mare Island Navy Yard, and the threat was not taken seriously.

However, that May, President Theodore Roosevelt arrived in Bremerton for a short tour of the Navy Yard. School children with bouquets of flowers lined one side of the city wharf and Foreman Colin A. Douglas presented Roosevelt with a gift from the men in the Yard,[10] a piece of steel which had been taken from the keel plate of OREGON, polished and engraved with the profile of OREGON and set in a case made of rosewood.

After the President visited the dry dock and took a hard

A third view of the Yard shows the saluting battery and imposing General Office Building on the left. To the right are (one behind the other) the Dispensary (former Civil Engineers' Office), Labor Board Office (Bldg. 56) and the start of the Yard's first hospital. On the hill are the Marine Reservation buildings. Excavation for Building 78 shows to the left of Building 66.

Kitsap County Historical Society

look at businesses in Bremerton, he departed for Seattle. Soon word came from Washington that unless all the saloons were closed the Yard would be shut down. All saloons were closed in Bremerton and Charleston. Later a compromise allowed a limited number of saloons to operate in each town, but restricting the number of saloons in the towns was a recurring problem.

Late in July 1905, the Yard received authorization to dock SS DAKOTA. She had been built for James J. Hill's Great Northern Steamship Company and left Seattle with a cargo of freight for the Orient. Outside Cape Flattery her propeller shaft malfunctioned, and DAKOTA returned to Seattle.

The Navy Yard had the only dock on the West Coast large enough to hold her. This was not the first commercial vessel in the dock but was by far the largest. When DAKOTA settled on the keel blocks, there were but three feet to spare fore and aft and her bow extended over the end of the dry dock.

Concern was expressed about the effect on the dry dock floor of the weight of DAKOTA. Although her cargo had been off-loaded to her sister ship, MINNESOTA, DAKOTA still weighed 21,000 tons against OREGON'S 13,000; but the dock passed the test.

> *Dockmaster George H. Trahey laid the keel blocks to such good purpose that each one took its share of the immense load as if moulded to the position.*[11]

The repair work was done by workmen from Moran Brothers, who struggled for three days to free the propeller hub from the shaft to determine the extent of the damage. A large force worked day and night to complete the ship in less than the predicted two weeks. Each day crowds of sightseers

More than 4,000 sightseers, fascinated by the 21,000-ton steamship DAKOTA, were drawn to the vessel's drydocking in August 1905. Steamers brought visitors from Seattle, Tacoma, Olympia, Fort Flagler, Everett and other outlying areas. DAKOTA was termed "the greatest sight ever witnessed in this Yard." University of Washington

Scenes like this were not unusual when the big battleships held open house. Visitors who came to view the Navy's pride were, in turn, looked over by sailors on the adjacent ship. Note the adventuresome ladies on the gun mounts, climbing the ladders and being ushered through a door by a courteous sailor. Puget Sound Naval Shipyard

thronged the dockside. On August 12, DAKOTA was undocked and returned to Seattle.

The Yard force cleaned and painted DAKOTA's hull and the Bremerton Searchlight commented that: "Had the work of repairing the ship been placed in the hands of our Navy Yard foreman the results would probably been more satisfactory to the company which is footing the bills."[12] SS MINNESOTA arrived in 1909 with a broken propeller shaft and this time the Yard did the repair work.

The need for a second drydock was apparent. The approaching completion of the canal through the Isthmus of Panama would increase the number of ships in the Pacific. The Puget Sound Navy Yard had the only dry dock big enough to handle the Navy's newest ships. With the increased amount of work coming to the Yard, it might be necessary to move a ship out of the lone dry dock to make room for a ship with a more urgent need.

Barclay said there was no room in the existing Yard for another dry dock. Assistant Secretary of the Navy Charles H. Darling favored a second dry dock, proposed many improvements and the purchase of more land. However, Congress refused to buy more land because of the high asking price.

Barclay's 62nd birthday brought him to the mandatory age for retirement in September 1905. On September 11, Burwell returned for his second tour as Commandant, this time for three years.

"No one could be assigned here as Barclay's relief who would be more welcome than Captain Burwell. Friends of the Yard have implicit faith in Burwell," stated the August 25, 1905, BREMERTON NEWS.

Burwell's first priority was to convince Navy officials the Yard did have room for another dry dock, and it would not be necessary to purchase more land.[13] The Congressional representatives of Washington State used all their powers of persuasion to achieve approval of the new dry dock; Secretary of the Navy Charles J. Bonaparte said it was an urgent necessity.

Burwell received the following telegram from Congressman Wesley L. Jones on March 27, 1906: "We have assurance from naval committee that they have provided for another dock. Shake."

Total cost of the dock was limited to $1,250,000. The first borings were made at the site west and a little south of the original dock, on July 19, but actual construction did not begin until 1909 when acceptable bids were received.

The Yard's size increased slightly in 1906. In order to protect the Yard's water supply from waste water released by a Chinese laundry in Bremerton, the Navy Department paid William and Sophia Bremer $4,000 for ten lots on the Yard's north boundary. This brought to 32 the number of lots purchased from the Bremers for the Yard's water system.[14]

The following year a gate was opened at Warren Avenue, the first on the north boundary of the Yard. With Bremerton expanding to the west, workmen appreciated the convenience of this gate, which was open at the beginning and end of each work day and during the lunch hour. Burwell initiated "comfort stations" in the Yard where workers could wash up and eat their lunches.

The year 1908 was exciting for the people of Puget Sound. Glistening with new coats of white paint, ships of the Pacific Fleet arrived in Seattle on April 8. The armored cruiser USS WASHINGTON (AC11), was making her first visit to her namesake state. Her wardroom silver, presented by Washington State at her launching in 1907, was displayed at a reception during the festivities of the Fleet's visit.

To answer queries regarding WASHINGTON's classification and the distinction between battleships and cruisers named for states, the April 12 SEATTLE TIMES explained:

. . . after battleships, which of course are armor plated and steel protected, there come two kinds of cruisers -one the steel armored cruiser and the other the unprotected cruiser. The battleships, as a rule, and the steel armored cruisers are named after the states while the unprotected cruisers are named after cities.

Before heading south for maneuvers, the ships visited Puget Sound cities and called at the Navy Yard for coal and repairs.

The SEATTLE POST INTELLIGENCER on April 16 announced the following ships were at the Navy Yard: NEBRASKA, WISCONSIN, OREGON, WHEELING, BOSTON, PRINCETON, ROWAN, GOLDSBOROUGH, CALIFORNIA, PENNSYLVANIA, COLORADO, ST. LOUIS, MILWAUKEE, SOUTH DAKOTA, SATURN, PHILADELPHIA and NIPSIC.

The Yard's lone dry dock experienced the busiest weeks of its existence with all departments rushing to complete work. Electric lights were installed on the piers to enable ship movements at night. The Yard's biggest payday occurred April ll; the paymaster toiled two full days to pay 2,100 men a total of $70,000 for two weeks' work.

The authorized work was completed a day ahead of schedule. On May 2, 1908, the eight armored cruisers of the Pacific Fleet returned PHILADELPHIA's 13 gun salute as they sailed away for San Francisco. Busier days were ahead. The Atlantic Fleet was coming also. Sixteen battleships and supporting vessels left Hampton Roads, Virginia December 16, 1907, passed through the Straits of Magellan and were now sailing up the west coast of the Americas.

On May 24, an estimated 400,000 people flocked to Seattle for the arrival of this fleet, now known as the Great White Fleet.[15] Once again the welcoming festivities included

Taken from the south shore of Sinclair Inlet, this westernmost part of another panorama shows the town of Charleston and the Navy Yard's coaling station. The two ships, one almost completely hidden by the other, are USS MARYLAND and USS SOUTH DAKOTA.

Kitsap County Historical Society

parades and gigantic fireworks displays. In addition, Grays Harbor sent 16 bear cubs. Each battleship had a live ''teddy-bear'' in addition to the other animal mascots the ships carried.

The fleet sailed for Tacoma on May 27 for further festivities. The four ships of the First Division, CONNECTICUT, KANSAS, VERMONT and LOUISIANA, left for San Francisco; the others proceded to the Navy Yard for repairs and supplies, including coal.

''For the convenience of visitors'' the May 23 BREMERTON NEWS reported the dates ships would be in the drydock; KEARSARGE — May 27, ILLINOIS — June 2, GEORGIA — June 6, NEW JERSEY — June 12, RHODE ISLAND — June 17, and VIRGINIA — June 22. The OHIO, MINNESOTA, MISSOURI, NEBRASKA and WISCONSIN also visited the Navy Yard.

The Great White Fleet assembled in San Francisco to continue its circumnavigation of the globe. Steam-powered steel battleships had never attempted such a long voyage, and many predicted the trip was technologically impractical. However the 46,000 mile tour was completed without a serious breakdown. The tour also showed off U.S. naval strength to discourage Japan's further expansion in Asia.[16]

Burwell's ability and leadership had been recognized by the Navy with a promotion to Rear Admiral on June 6, 1906. By 1907 he was ninth highest of the Navy's 22 Rear Admirals and only Admiral George Dewey, hero of the Battle of Manila, held rank higher than Rear Admiral.

The next year Bremerton and Charleston laid aside their rivalry and gave Burwell's name to a new street which connected Third Street in Bremerton with Seventh and Cedar

Farther east, USS MILWAUKEE, then in reserve at the Navy Yard, is anchored off shore of the Navy Yard's two brick storage buildings. USS WASHINGTON is centered in the picture and USS TENNESSEE partly conceals the Receiving Ship PHILADELPHIA. The officers' quarters are visible on the ridge at the right; the snow-capped Olympic Mountains fill the skyline.

Kitsap County Historical Society

Smartly uniformed band members pose in front of the wooden Marine Barracks. When the Marines moved to the west end of the Navy Yard, this building, much enlarged by then, was remodeled to serve as a Naval Disciplinary Barracks. Puget Sound Naval Shipyard

The third picture shows the Navy Yard's industrial area and the town of Bremerton. USS WISCONSIN is anchored nearest the shore on the left; USS CALIFORNIA is refueling from a coaling barge in the stream. In the center is USS WEST VIRGINIA; behind her are USS COLORADO and USS PENNSYLVANIA. With the exception of the smaller cruiser MILWAUKEE and the battleship WISCONSIN, all the ships in these three pictures were armored cruisers of the Pacific Fleet's First Squadron. Puget Sound Naval Shipyard

Rear Admiral William T. Burwell posed with his staff on the steps of Building 50. This picture was taken after Burwell attained his flag rank in June 1906 and before November 1906 when Major L. H. Moses, U.S. Marine Corps, was detached. Moses is on the step above and slightly to the right of Burwell. Commander Vincenden L. Cottman, second in command, is in the front row on the right.
Kitsap County Historical Society

Streets in Charleston. The name Burwell Park was given to a children's playground established on the approximately one-and-a-half acres of government land at the south end of Dock Street (now Park Avenue) between Burwell Street and Fourth Street in Bremerton.[17]

Burwell had long been a strong supporter of the area. Each year he invited school children and teachers of Kitsap County to the Yard on his birthday (July 19), pointing out that one could never begin too early to teach patriotism. The children toured the ships and enjoyed picnics on the hill while the band played in their honor.

The band, organized during Burwell's first tour as Commandant, played in residential and industrial areas in the Yard and in nearby towns. Local boys borrowed instruments and practiced with the band.[18]

Burwell's varied interests included raising various kinds of fowl in the gardens behind his quarters. One of his last acts before retiring was to release his flock of Mongolian pheasants, thereby providing breeding stock for the pheasants still living in the non-industrial area of the Puget Sound Naval Shipyard.

Burwell turned over his command to Rear Admiral John A. Rodgers on July 18, 1908. Following a simple ceremony in his office, Burwell and his wife said their goodbyes at the pier before boarding the tug for Seattle. The Burwells planned to make their home in Seattle after touring Europe.

Under Rodgers, work began on the second drydock. Impressive ceremonies marked the groundbreaking on a bitterly cold January 4 morning in 1909. Miss Myrtle Jones, granddaughter of Dexter Horton of Seattle, pressed a lever on the contractor's steam shovel. Its bucket plunged into the ground and came up with a cubic yard of earth, as the Navy band played patriotic airs. People cheered and every whistle in the Yard blew.

Workers pause for the photographer inside the Construction and Repair Department's Drafting Room in October 1908. Reading clockwise, starting with the man in the foreground are: Herbert Burton, Justin Comment, Frank Cullen, Zip Diamond, J. C. Foley, Chief H. E. Bailey, John Heck, Edward Crawford, Arthur Holden, Harry Green, Harry Holt, and Emil Klinghammer. Arthur Holden (to whom the arrow points) was one of the first apprentices in the Navy Yard.

Kitsap County Historical Society

Contractor C. J. Erickson of Seattle built the new dock 400 feet west of the first dry dock. Earth from its excavation and the grading of the adjacent hill was moved to the west end of the Yard to fill in a swamp in preparation for a Marine Parade Field.

While under construction, Dry Dock 2 sometimes served as a movie theater. Men of the fleet, invited by the crew of USS COLORADO, watched movies of the Moran-Nelson championship fight projected on a large canvas stretched across an end of the dock site. Cheers of the audience could be heard for miles.

In October 1909, the Yard began a new system of transporting supplies from Seattle, ferrying freight cars to the Yard and then moving them on tracks to the various shops. The next February the first load of stone for the new dry dock arrived on a barge carrying nine 40- or 50-ton rail-car loads of granite blocks from Index, Washington. It was estimated that 50 stone workers would work a year at the site shaping 200,000 cubic feet of granite to line the dock.

The beginning of 1909 was a very cold winter. A chill was also felt by many who worked in the Yard as a major organizational change took place in all Navy Yards. Many feared they would be demoted or lose their jobs.

President Roosevelt insisted on the Navy adopting a more efficient bureau system. At that time each Navy Bureau was autonomous in each Navy Yard, with its own crew of men, shops and appropriated funds. In the closing days of Roosevelt's administration, the Newberry Plan[19] went into effect to stop the duplication and increase cooperation.

With the new system, the Departments of Equipment, Steam Engineering, Ordnance and Public Works were consolidated under the head of the Construction and Repair Department. The former heads of each department became Inspectors of their discipline and reported to the Naval Constructor.

The Departments of Supplies and Accounts and Medicine and Surgery were not affected; the change affected only industrial functions. The Ordnance Officer was still in charge of the gun shed on the waterfront near the ravine and the Magazine,[20] which was being built on Ostrich Bay. The Equipment Officer was responsible for the coaling plant in the west end as well as the wireless station which had been started on the hill above the officers' quarters.[21]

Early in the administration of President William Howard Taft, the new Secretary of the Navy, George von L.

Construction for Dry Dock 2 started in 1909 and forced the moving of Building 50 from the edge of the site. The hospital, completed in 1903, shows in upper left. Buildings 78 and 104 appear on the right. Kitsap County Historical Society

Meyer, announced more changes. Now the Naval Constructor's authority would be limited and he would report to the Commandant, who would be the sole representative of the Navy Department in each Yard.

The new name of Consolidated Manufacturing Department was kept, but repair work was divided into Hull and Machinery Divisions. At Puget Sound, Commander A. H. Robertson of the Machinery Division had control of the machine shop, boiler shop, foundry, pattern shop, electrical shop and power plant. Naval Constructor J. D. Beuret was in charge of the Hull Division and its shipfitters, sheetmetal workers, joiners, riggers, laborers lobby, blacksmith, boat, paint and sail shops. Later, a separate Public Works Department was established.

Early in February 1910, offices for the Naval Constructor were set up on the second floor of the Equipment Building. Inspectors and draftsmen moved into its sail loft. All administrative offices were to be moved from Building 50, which was dangerously close to the crumbling bank caused by excavation for the dry dock. The former General Office Building was moved to the beach below the officers' quarters.

Shortly before the expected return of the Burwells to Seattle, the January 5, 1910, newspapers carried the sad news of Admiral Burwell's death in Wales after a short illness. The Navy Yard towns mourned the loss of their "staunchest supporter".[22]

On July 10, 1910, Rear Admiral Rodgers' flag was lowered from PHILADELPHIA and Captain Vincenden L. Cottman became the new commandant.[23] That same year, Cottman became a Rear Admiral and the Commandant of the 13th Naval District, which until then had been under the cognizance of the Commandant of the 12th Naval District.[24]

Excavation for the new dry dock was finished August 16, 1910, and Contractor Erickson's men began laying its foundation. The ceremony of the laying of the final stone took place April 27, 1912. The program for the event compared the statistics for the Yard's two dry docks. "In an emergency" the first dry dock could dock a ship drawing 29 feet, 10½ inches, while the new dock could accept one drawing

The massiveness of Dry Dock 2's construction is evident in this December 4, 1911 photo. Its great weight of concrete made it unnecessary to install a pumping system under the dock to relieve ground water pressure to prevent an emptied dock from lifting.
Puget Sound Naval Shipyard

almost 38 feet. It also noted that the new dock would hold 30,860,000 gallons of water.

After the ceremony, many months of finishing work remained. The pumping plant and pipes were installed and entrance of the dry dock prepared so the caisson could be put in place. The caisson, built by the Seattle Construction and Dry Dock Company, formerly Moran Brothers, measured 127 feet long by 47 feet high. Complete with engines, pumps and other gear, it cost $125,000. Eleven men rode the wildly pitching caisson down the rails at its launching on January 3, 1912. It was in place the following January.

Dry Dock 2 was dedicated on a beautiful March 1, 1913. Extra boats were added to the regular runs from Seattle and Tacoma to carry the crowds wishing to attend. A steamship brought more than 300 people from Olympia, including Governor Ernest Lister, who gave the principal address. On the arrival of the Governor's party, an estimated 3,000 watched as the signal was given for the USS OREGON to enter the dry dock.

OREGON, which had almost filled Dry Dock 1, took up only a part of the new 827-foot long, 145-foot wide, 38-foot deep dock.[25] Following the dedication ceremonies, at which 500 school children sang patriotic airs, 800 guests were entertained on the ship. Others toured the shops, while the official party was served luncheon at the Washington Veterans Home across Sinclair Inlet.[26]

The increase of industrial facilities at the east end of the Navy Yard crowded the hospital and the Marine reservation. The idea of purchasing more land on which to build a hospital was discussed and discarded, and a site on the high ground west of the ravine was selected.

The hospital, consisting of three brick buildings connected by solaria, opened January 1, 1912. Surgeon F. C. Cook, Medical Officer in Command, was assisted by two doctors, a pharmacist and 18 enlisted attendants.

The old naval hospital building was sold to private interests and moved through the Warren Avenue gate in 1912 to 7th and Chester where it served as a community hospital for many years. It was the first Harrison Hospital and later the Horton Nursing home.

The Navy transferred 40 acres of land at the northwest corner of the Navy Yard to the Marines.[27] Lieutenant Col-

Yard personnel and visitors welcome an old friend, USS OREGON, as she participates in the opening of Dry Dock 2 on March 2, 1913.

Kitsap County Historical Society

Photographed from the south in the otherwise empty dry dock, OREGON dramatically portrays the size of Dry Dock 2, in comparison with Dry Dock 1, as pictured in Chapter One.

Kitsap County Historical Society

onel Joseph Pendleton,[28] remembering how quickly men and officers outgrew the 1899 barracks and officers' quarters, planned a barracks building suitable for 500 men and quarters for three officers. Pendleton's brother-in-law, Colonel Charles Doyen served as Marine Commandant when quarters were completed in 1912 on the street that now bears his name. The brick barracks, later known throughout the Corps as "Old Red", was accepted from the contractor May 30, 1912. Bachelor officers' quarters (Qtrs M4) and a stable were soon added.

With the 235 Marines moved to the west end, the old barracks became a disciplinary school for military offense prisoners. This was the second school of this type in the U.S. for Navy and Marine Corps enlisted men. Its staff consisted of a first sergeant who was assisted by a sergeant, four corporals, 17 privates, a drummer and a trumpeter. School opened with seven marines and ten sailors, but with an expected capacity of 60 prisoners.

More serious cases were sent to the Prison Ship. In order to relieve prison congestion at Mare Island, NIPSIC became a prison ship, when the larger USS PHILADELPHIA became the Navy Yard's Receiving Ship in 1903. Ray Raines described the NIPSIC thus:

". . . all portholes and windows were barred over, and she was painted black from waterline to her stubby mastheads. The prisoners served court martial sentences for serious crimes — not just drinking or a few days of AWOL. They were serving time at hard labor, and did menial work around the Yard, such as cleaning streets, raking leaves etc. They wore prison-gray uniforms and worked under the watchful eyes of a Marine guard armed with a sawed-off pump-action shotgun — loaded with buckshot . . . I was told at the time, and quite likely it was true, that if a prisoner in a work party escaped, the Marine guard would be court martialed, and if found guilty of neglect, he would be sentenced to serve out the unexpired term of the escapee . . . I never heard of a prisoner escaping from a work party."

The cruiser USS CHARLESTON (CA22) became the Receiving Ship in 1912. PHILADELPHIA then became the Prison Ship.

During this period the Navy disposed of its last wooden ships and in spite of efforts to save NIPSIC because of her historical significance, she was stricken from the Naval Register December 11, 1912. Sold at auction to George Willey of Seattle for $7,375, she was used as a cannery tender. NIPSIC's career ended in July 1915, when, because of the war in Europe, metals were in high demand. She was

View from the hills to the west of the Navy Yard in 1915 overlooked homes in Charleston in the foreground, the new Marine Barracks and Officers' Quarters in the center and the hospital on the rise behind and slightly to the right of the Barracks.

Puget Sound Naval Shipyard

NIPSIC, the original Station Ship, added one more function to her colorful past in 1903. She became the Yard's Prison Ship when USS PHILADELPHIA became the Receiving Ship. Puget Sound Naval Shipyard

Detentioners practice rowing on Sinclair Inlet alongside USS PHILADELPHIA. Detentioners were new recruits who were quartered separately for a few weeks after arrival in order to prevent spread of disease. Puget Sound Naval Shipyard

The construction of the Central Power Plant, authorized in 1907, is under way here in December 1909 east of Dry Dock 1. This plant took over the work of the separate power plants located throughout the Navy Yard. It furnished power for pumping the two dry docks, the Yard's heating and lighting system, water and fire protection systems, and compressed air. Puget Sound Naval Shipyard

Hot gases rise through the exhaust vents of two Schwartz metal melting furnaces in the foundry. This 1914 photo shows a melt in process, preparatory to the pour. The operator turns the large wheel, on which his hand is resting, to tilt the furnace and start the pour. Puget Sound Naval Shipyard

Rear Admiral Vincenden Cottman poses with his department heads in February 1914. Front row, left to right: Colonel J. H. Pendleton, Commanding Officer Marine Barracks; Commandant Cottman; Commander D. W. Blamer, Captain of the Yard; and Commander J. R. Brady, Engineer Officer. Second row: Surgeon F. C. Cook, Commanding Officer Naval Hospital; Commander F. A. Traut. Commanding Receiving Ship; Constructor F. S. Smith and Paymaster George Brown. Third row: Civil Engineer Luther E. Gregory; Paymaster R. Nicholson; Lieutenant C. L. Arnold, Inspection Officer; and Surgeon J. W. Backus, Dispensary. Upper row: Captain J. F. McGill, Commanding Disciplinary Barracks; Paymaster H. H. Alkire, Accounting Officer and Lieutenant M. F. Draemel, Aide to the Commandant. Puget Sound Naval Shipyard

burned at Lummi Island for the metal in her hull, especially the copper plates from her bottom.

A major advancement in the Yard's operation occurred in May 1911 when the new Central Power Plant went into operation. Building 106 provided power for the Yard's lights, electrical machinery, hydraulic power for compressed air machinery and steam to heat the buildings. William Bankhead, the Yard's chief electrician, became foreman of the plant. This position, paying $6.56 a day, was considered the top civilian post in the Yard.

The February 24, 1911, Bremerton Searchlight reported on the new storehouse (Bldg. 157) under construction in the Yard:

> *. . . the largest building ever constructed by the Navy Department at any yard . . . It is a magnificent building of reinforced concrete, four stories high, with basement, and will contain four and a half acres of floor space . . . There is a large court in the center of the building and into this will run double railroad tracks from pier 8 . . . The pier will be used exclusively for the landing of stores from ships to the general storehouse.*[29]

PSNY received a 600-ton anchor chain testing machine in January 1912, thus becoming the second Navy Yard in the United States where chains could be tested. The machine, run by a hydraulic engine, was capable of testing a chain length of 240 feet. The equipment came from the Boston Navy Yard, which had received a new machine for testing larger chains because the Navy's new battleships were being built on the East Coast.

As the time for Cottman's retirement approached, there was much speculation as to who would be the next commandant. However, no successor was named, and the Captain of

View looking east from the Radio Station in 1915 shows the graded top of the hill on which the Marines built their parade field in 1899. East of the Marine Quarters and Barracks (hidden by trees) is the foundry. South of the foundry is Building 109, the Erecting and Boiler Shop. Across Main Street (now Farragut Avenue) is Building 78.

Puget Sound Naval Shipyard

Men and horses furnish the power to clear the vegetation and excavate the ground for fuel oil tanks in the ravine in 1915. Two large brick buildings, on either side of the mouth of the ravine, are provision and clothing storehouses, originally built for ordnance storage.

Puget Sound Naval Shipyard

Major buildings in the background of this 1917 view of the new building ways, are the Central Power Plant (Bldg. 106) and the Smithery and Shipfitters Shop (Bldg. 84) Puget Sound Naval Shipyard

the Yard, Commander D. W. Blamer took over as Acting-Commandant on February 14, 1914.

Later in 1914 the Navy Department expanded its holdings in Kitsap County. An 82-acre Liberty Bay site, for the long discussed torpedo station, was turned over to Blamer on July 21.[30]

Although there had been talk of building a new and larger coaling station in the west end of the Yard, a sign of the times was the report in the Bremerton News on September 25, 1914:

> *. . . The coming of the battleship fleet, the dreadnaughts, of which several are oil burners, makes it absolutely necessary for the Yard to be equipped with sufficient oil. The difficulty of securing sufficient coal on this coast for the fleet also makes the burning of oil a great necessity.*

Five steel oil storage tanks were authorized and work began on the first two in 1915. These were placed in the ravine with a dam downstream of each to provide containment in the event of an oil spill.

In July 1915 work began on a new high-powered wireless station that would put the Yard in contact with stations on the Pacific Coast, Hawaii and possibly the Atlantic coast. Keyport was considered as a location, but instead improvements were made at the small station on the hill above the quarters, which Ray Raines described thus:

> *The instruments were housed in a white cottage-type building. (Bldg. 133). There was a Chief Petty Officer in charge, and there were several lower ratings to stand watches. They all lived in the station building. What we now call radio towers were, in those days, called "wireless masts". The people responsible for the construction of the station had selected two, very tall, properly spaced living fir trees, and these became the masts. They were left standing on their own roots; but all bark and limbs were removed. They were then guyed to the ground by cables, and finally painted white. The antennae consisted of four strands of wire, held at proper spacing by a spreader at each end.*

Improvements in 1915 included new metal masts and improved equipment.

At last word arrived of the naming of the new commandant, Captain Robert E. Coontz,[31] who took command on July 20, 1915. The 54-year-old Coontz chose Puget Sound when given his choice of three yards and his appointment was well received in the area, even though people had hoped for a Rear Admiral.

For years the Yard had asked to be assigned major ship construction in order to extend work over a long period, thus preventing layoffs when repair jobs were completed. With the outbreak of war in Europe, our Navy's needs became apparent and the Yard received $20,000 to construct building ways large enough to hold major ships. Construction of the ways began July 2, 1915.

No ceremony marked the occasion, but it was an omen of the new role the Yard would play during Coontz' administration.

Rear Admiral Robert E. Coontz stands in front of USS WYOMING in 1918, between his orderlies, Mate Jerry Blake (former warden on NIPSIC) and Marine Sergeant John Delaney. During World War I, no badge was valid unless a hand of the esteemed Mate appeared on the left shoulder of the photographed worker. Gertrude McGowan Madden

Chapter Three

Growth for a New Mission

1915-1928

When Captain Robert E. Coontz arrived in July 1915, the Puget Sound Navy Yard was about to benefit from the largest naval shore establishment development of this country's history. The need for expansion related directly to the broadening of "The Great War" in Europe and to a lesser degree to the unrest at the Mexican border because of Mexico's civil war.

In spite of President Woodrow Wilson's efforts to keep the country out of war and the desires of the populace to remain neutral, it appeared the United States would become involved in Europe. The sinking of the British Cunard Liner LUSITANIA in May 1915 and the French liner SUSSEX later that same year spurred Congress to appropriate funds for nationwide Yard and Station development; $60,000 was allocated to the Puget Sound Navy Yard. In addition to the already mentioned $20,000 for ship building ways, $25,000 was allocated for enlarging the railroad system and $15,000 to continue dredging around wharves at the second drydock.[1]

The President called for an increase in the armed forces. A bitter discussion in Congress led to the appointment of the Helms Commission and the establishment of the Board for the Development of Navy Yard Plans.[2]

The Commission, under the leadership of Rear Admiral J. M. Helm, stated that the major objective of the Navy was the protection of commerce and dealt specifically with a readiness study of Pacific Coast bases and their ability to support fleet operations. The report recognized the West Coast was too far from the war zone to handle battle repairs. However, western yards could specialize in new construction with repair and overhaul as a secondary consideration. The report stressed the need to develop ship-building capabilities at Puget Sound and recommended an appropriation of $2,500,000 for a third dry dock and necessary piers, grading and filling.

The Board for the Development of Navy Yard Plans asked for a comprehensive plan from each of the Navy's Yards and Stations. The Navy Yard at Puget Sound was ready. The Public Works Officer, Captain Luther E. Gregory, and other department heads had prepared studies and back-up reports.

There had been talk of the need for a third dock for many years. Draftsman Victor Hulteen of the Public Works Department suggested building a shallow dry dock in which ships could be built. This would make materials more accessible to workmen, alleviate some difficulties during launching and cost no more to erect than building ways. The idea was new in the United States; however, Germany, France and England had successfully used smaller versions of such a dry dock for several years.

Gregory, stressing the cost advantages of the structure, forcefully presented the plan to the Development Board in March 1917. The plan was accepted. Sound Construction and Engineering Company received the contract to build the dry dock and by August of that year work began east of the first dry dock. With this shipbuilding dock, the Puget Sound

As the Navy Yard developed, laying railroad tracks on Main Street (later Farragut Avenue) continued in 1917. Just beyond Building 78, on the immediate right is Dry Dock 1. Before the Central Power Plant was completed, older buildings had their own heating plants. The chimney in the foreground corner of Building 78 marks that building's furnace room.
Puget Sound Naval Shipyard

A 150-foot-boom crane was used in construction of Dry Dock 3. The excavcation and sheet piling installation near completion in this June 1918 photo. Buildings in the background (from left) are: Shipfitter and Boiler Shop (Bldg. 178), Metal Storage (Bldg. 107), Plate Metal Shop (Bldg. 102) and, still under construction, General Storehouse (Bldg. 290). To the east of the dry dock site is the Joiner Shop (Bldg. 91).
Puget Sound Naval Shipyard

Fitted with a dump-box in place of the trunk, this 1914 Model T, with cables around rear tires for traction, became the Navy Yard's first dump truck. Listed as No. 1 Refuse Remover, this Ford and two carts drawn by mules hauled the Yard's rubbish. According to this 1917 photo, some trash was dumped at the water's edge. Puget Sound Naval Shipyard

Navy Yard took the lead in an improvement which would be adopted later by naval and civilian shipyards.[3]

A major change taking place in the Navy Yard at this time was the substitution of automotive for horse power. Two electric trucks did not prove practical, but Henry Ford's Model T made its appearance in the Yard in 1914 and was in use for many years. A Packard truck and a one-ton ambulance appeared soon afterward.

In 1917 a gray Winton "6" automobile was purchased to replace the horses and carriage then in use by the Commandant. Frank Batson, who had cared for the horses and equipment, became the chauffeur. In an interview in SALUTE, Bertha Coontz Kokko, who had enjoyed four years of her childhood in the Navy Yard as daughter of the Commandant, recalled the transition from hay to gasoline:

> *When the car replaced the Victoria and horses, Batson, then in his fifties, had to learn to drive. The first time he took us for a ride he was so nervous he fainted. However, he finally mastered "the new-fangled thing", but he never really trusted it as he had Bob and Sam. His happiest day, after the Winton came, was the day he turned the keys over to George Fisher who fell in love with the car and served it well.*[4]

Early in 1917 the country was inflamed by news of an intercepted German Foreign Office cable (the Zimmerman Letter) offering Mexico most of the southwestern territory of the United States for a military alliance if the U.S. entered the European war. The U.S. declared war on Germany on April 6, 1917, after that country began unrestricted submarine warfare despite promises to refrain from attacking our ships.

The next day Coontz directed the Navy yard to work the following Sunday ". . . with no officer relieved of duty without direct permission of the Commandant". The Yard was off and running; constructing, converting, modernizing, improvising and building.

As a neutral nation, our policy had been to remove from our territorial waters all ships belonging to combative nations. Two German ships, the steamer SAXONIA and sailing vessel STEINBECK, were captured and laid up near Port Blakely. The crews remained aboard under observation until replaced by U.S. military personnel. Within a week of the declaration of war, the German ship PRINZ WALDEMAR was commandeered, overhauled at the Yard and turned over to the Shipping Board in Seattle.

The United States entry into the war triggered wholesale enlistments. Within ten days, approximately 700 men of the Oregon and Washington Militia mobilized at Puget Sound;

Below the Marine stable (upper right), which later became the Chiefs' Club, land had been cleared by June 1917 for the barracks site extending east to the Marine parade field. Many of the pupils attending the school house on the hill would eventually be employed in the Yard.
Puget Sound Naval Shipyard

new recruits swelled their number. The men lived in tents until October 16, 1917, when those still at the camp were moved into barracks buildings erected west of the Marine compound. The Surgeon General of the Navy reported the next year that:

> *Eight separate units at present constitute the camp proper, each unit being complete in itself, as far as sleeping quarters, kitchen, mess hall, wash rooms and toilets are concerned.*[5]

He also reported that eleven 100-watt Mazda drop lamps, arranged in the center of the 150 by 18 foot buildings, furnished "ample" light. The number of men in each building was limited to 50.

As men in the work force left for military duty, school girls and housewives took their places. A 1914 survey showed only six women employed at the Yard, three seamstresses and three clerks, but during the war many women would be at work in the Yard. The Commandant spoke to the senior class of Union High School,[6] telling the young people of their patriotic duty to work in the Yard. Almost the entire class applied for jobs.

Secretary of the Navy Josephus Daniels startled the military services in March 1917 by decreeing the Navy would enlist women as Yeomen (F) in order to free enlisted men in the event of war. Over 11,000 women, affectionately known as "Yeomanettes", enlisted. More than 200 served in clerical positions in the Thirteenth Naval District.[7]

In his book FROM THE MISSISSIPPI TO THE SEA, Coontz wrote that the Yeomanettes:

> *. . . organized several companies of infantry . . . became experts in the handling of arms, in firing and target practice, and in rowing and managing cutters and whale boats . . . If the war lasted a few months longer I feel positive that they would have been accepted for service and would have gone to France.*[8]

He and Mrs. Coontz treated the young women as part of the family, chaperoning them on private as well as official outings. However, some people had difficulty accepting the idea of women in military service.

The Surgeon General of the Navy reported in 1918:

> *There are at present attached to the navy yard in connection with the Thirteenth naval district 101 enlisted females. These girls are living in houses situated in Bremerton and Seattle, Wash. In some instances the hygienic conditions under which they exist*

An instructor trains women to operate engraving machines. Each machine received power from an individual motor in contrast to other similar shops where one motor served many machines. Puget Sound Naval Shipyard

Members of the Yeoman(F) rowing team pose with their coach, a Chief sporting eight "hash marks". Standing to his right is Lyle McCloud (Sauer), for whom the Girl Scout Camp north of Belfair was named. The young women are attired in the prevailing sports costumes of the early decades of the century, middy blouses and sateen bloomers. Lettie Gavin

These World War I rivet passers and rivet catchers foreshadowed "Rosie the Riveter" of World War II fame. They relax here, holding their tongs and rivet buckets, in front of an example of the trade. Rivets in the ship's hull are visible behind the women on the left side of the picture.

Puget Sound Naval Shipyard

are bad, particularly with respect to ventilation, which is defective, due to overcrowding and improper and insufficient bathing facilities.

Often two girls occupy one small poorly ventilated room, and of late quite a number have come down with mumps, measles and tonsillitis. Under present conditions it is not possible for the medical department charged with the responsibility and care of the health of the enlisted woman to do much toward disease prevention.

If the enlisted woman has come to stay, it appears reasonable that consideration be given to this question. I have to propose the erection of a building, in the nature of a barracks, on the naval reservation for the purpose of housing female enlisted personnel. It is realized that this suggestion is fraught with novel possibilities, but it is believed that the best interests of the service demand more careful regulation of the living conditions and movements of this new naval adjunct.[9]

Life did not seem grim to the young women themselves. Helen McCrary Burns, who had been in charge of Yeomanettes at the Yard, told Lettie Gavin,[10] "Oh, we had so much fun." Gertrude McGowan Madden,[11] who enlisted when she became 18, echoed, ". . . it was a sad time for so many but I had a wonderful time. I have such wonderful memories."

As the number of personnel, military and civilian, increased in the Thirteenth Naval District, the work load at the hospital in the Navy Yard tripled and quadrupled. Twenty-five tents housed the overflow of patients until September 1917 when the first five of ten new pavilion wards were completed.

The arrival of 15 members of the Navy's female Nurse Corps [12] in June 1917 greatly increased the efficiency of the hospital staff. There were no quarters available for them in the Yard and they were housed in Charleston above a "moving picture house and billiard hall".[13] Their primary job was to supervise the hospital corpsmen.

In November 1917, Red Cross Nurse Julia Hippe [14] wrote to friends describing her experiences when she reported at the Navy Yard for duty. Having finally convinced the guards at the gate that she was authorized for entry into the Navy Yard, she was startled when directed to an ambulance. This was the vehicle used to transport the nurses to and from their quarters a mile away.

When influenza became epidemic in the Puget Sound area in the fall of 1918, twelve more Red Cross nurses were loaned to the hospital, and Yeomanettes requested and received permission to help the overtaxed nursing staff. There were 2,937 admissions during the epidemic. When the hospital could hold no more patients, the Armory and 48 hospital tents were pressed into service.

Yeomanettes who volunteered for additional duty could join the drill team and learn to operate light weapons. Here, those who were assigned to the hospital, are shown with the Drill Instructor in 1918. Mary Margaret Fitzgerald stands third from the left in the front row.

Lettie Gavin

AMBULANCE
U.S. NAVAL
HOSPITAL
PUGET SOUND WASH.

One of the first ambulances, possibly the first gasoline-powered one in the Yard, is seen in the industrial area in July 1915.

Puget Sound Naval Shipyard

Victory in 1918! "The war to end all wars" is over! That day church bells rang and ships' whistles blew continuously. The Navy Yard band stirred the crowds to dance on the streets. Signs and banners appeared everywhere. On Pacific Avenue, crowds surrounded the band and a formation of Yeomanettes. Puget Sound Naval Shipyard

In an effort to contain the spread of the disease, public gatherings were cancelled. Cloth masks had to be worn in public, leading to the expresssion "In Gauze We Trust", while servicemen complained, "We opened the tent flap, in Flu enza, out flew liberty."

At the Navy Yard most of the deaths from the disease were in a draft of infected recruits from the East. One doctor, one female nurse and two hospital corpsmen also died.

Nineteen-year-old Margaret Mary Fitzgerald,[15] stationed at the hospital, wrote to her parents on October 8, 1918:

> *The influenza is raging pretty bad here. Forty-six have died at our hospital within the last two weeks, and one of our Doctors died this morning with it. The movies, barber shops, churches, etc., are closed in Bremerton.*

Throughout the country, half a million Americans died in little more than ten months. American deaths from influenza were greater than the number of servicemen killed in any war; including the Civil War when 498,000 died in four years.[16]

The spread of disease was exacerbated by the crowded living conditions; the influx of civilians and military created a serious housing shortage in area towns. Uncomfortable living quarters, often make-shift basement apartments, persuaded many to endure the hour-long commute to Seattle.

The government instituted a liberal building program. A 45-unit apartment house (South Court) at Warren Avenue and Seventh Street, and 250 individual houses were built throughout the Bremerton and Charleston areas. Work also began on a 350-room hotel adjacent to the Yard, on land formerly used as Burwell Park. The hotel was not completed before the need for its service disappeared at the end of the War. Known as the Navy Yard hotel, it served various purposes before becoming Craven Center.[17]

In 1918, the Navy Yard received a 6-ton electric furnace which enabled it to make steel castings and avoid long delays in obtaining them elsewhere. The furnace was also used in the Navy's drive to re-cycle as much material as possible. It was estimated that at Puget Sound 200 tons of scrap steel could be salvaged with a savings of $94,000 a year.[18]

During World War I, PSNY acquired additional storage space. A nine-story concrete building (Bldg. 290) was erected just inside the Bremerton gate. Before the war the Yard's storage area was 134,424 sq. ft. By July 1918 it was 422,224, an increase of 215 percent.

The total area of the Navy Yard increased 25.74 acres. A small piece was obtained from the U.S. Housing Corporation to fill in a gap in the land owned at Park Avenue between Burwell and Fourth. Two pieces were bought in Bremerton. A 4.33-acre piece at the east end of the Yard was purchased for $22,330, and a 1.35-acre piece to fill out the northeast corner cost $4,620. A 20.05-acre piece squared off the Yard's northwest corner at a cost of $21,600. Most of the land was bought from the Bremer and Jertson families. [19]

The work force swelled from 1,500 in early 1916 to 6,500 in late 1918. The Machinery Division (Steam Engineering Department), which had 775 employees on June 30, 1917, increased to 1,561 civilians and 27 military in one year.

In the Commandant's Office alone, the number of personnel increased from 20 to 65. In his autobiography, Coontz wrote " . . . we constructed a connection between two old buildings" (Bldgs. 78 and 104) "and I used the new section for my headquarters — one wing for District — one for Navy Yard work."[20]

Two months before the end of the war, Coontz was transferred to Washington, D.C., where a year later he became the Navy's second Chief of Naval Operations. On Coontz's recommendation, the Naval District offices were moved to Seattle, when he was detached from the Navy Yard.

The war ended on November 11, 1918. After an initial drop to a little over 6,000, the work force increased to more than 6,600 at the end of 1919 in order to complete contracted work.

By war's end, the Yard had completed, or nearly completed, 1,700 small boats, 25 sub-chasers, seven seagoing tugs, seven submarines, two mine sweepers, and two ammunition ships.

Six 151-foot-long submarines were intended for the Imperial Russian Navy. Holland Electric Boat Company of New York had an order to build 11 boats and deliver them to St. Petersburg. Five were delivered before the Germans blockaded the Baltic Sea in 1916; the Russians requested the remaining disassembled submarines be shipped via Vancouver, B.C., and Vladivostok, over the new Trans-Siberian Railway to St. Petersburg. Before the parts left Vancouver, the government had been overthrown. The builder repossessed the boats and in 1918 sold them to the U.S. They were shipped to PSNY to be assembled and commissioned.[21]

The Navy Yard also built another submarine, O-2. The keel was laid August 1917, and the boat was launched in May 1918. She was 172-feet four inches long and 18 feet across with one three-inch gun and four 18-inch torpedo tubes.

The two ammunition ships were christened on December 16, 1919. Because of the need for these ships, the contractor for the new dry dock had quickly constructed the floor and a short section of the side walls and installed traveling cranes, so work could proceed on the ships during the dry dock's construction.

Neither rain nor chill discouraged the umbrella-protected audience from witnessing the launching of submarine O-2, May 24, 1918. More submarines, under construction on the building ways, are seen in the background.

Puget Sound Naval Shipyard

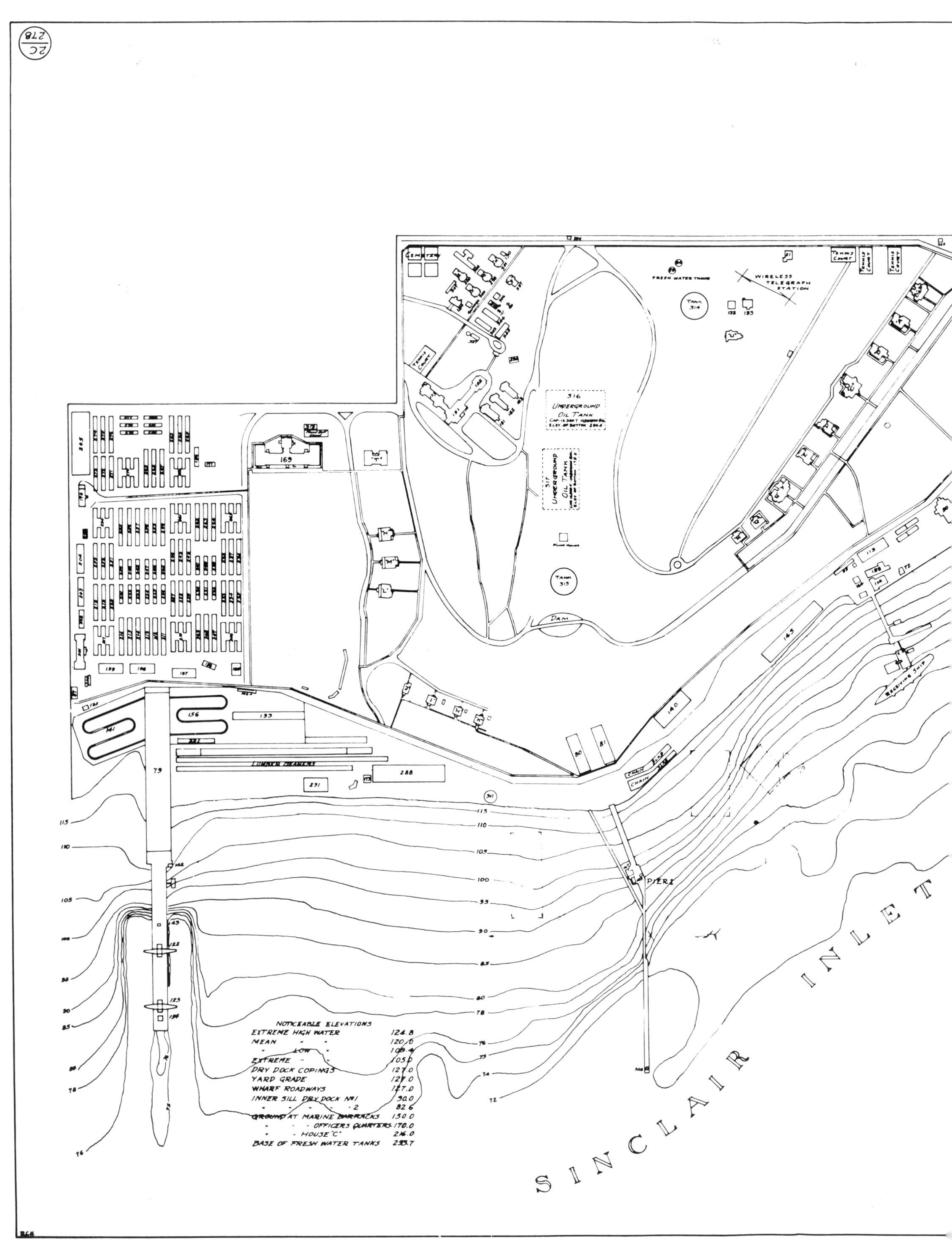
CEMETERY
FRESH WATER TANKS
WIRELESS TELEGRAPH STATION
TANK 314
TENNIS COURT
316 UNDERGROUND OIL TANK
317 UNDERGROUND OIL TANK
PUMP HOUSE
TANK 315
DAM
169
LUMBER TRAMWAYS
288
PIER 1
CHAIN
RECEIVING SHIP
NOTICEABLE ELEVATIONS
EXTREME HIGH WATER 124.8
MEAN " " 120.0
" LOW " 108.4
EXTREME " " 105.0
DRY DOCK COPINGS 127.0
YARD GRADE 127.0
WHARF ROADWAYS 127.0
INNER SILL DRY DOCK No 1 90.0
" " " " 2 82.6
GROUND AT MARINE BARRACKS 150.0
" " " OFFICERS QUARTERS 170.0
" " HOUSE "C" 216.0
BASE OF FRESH WATER TANKS 253.7
SINCLAIR INLET

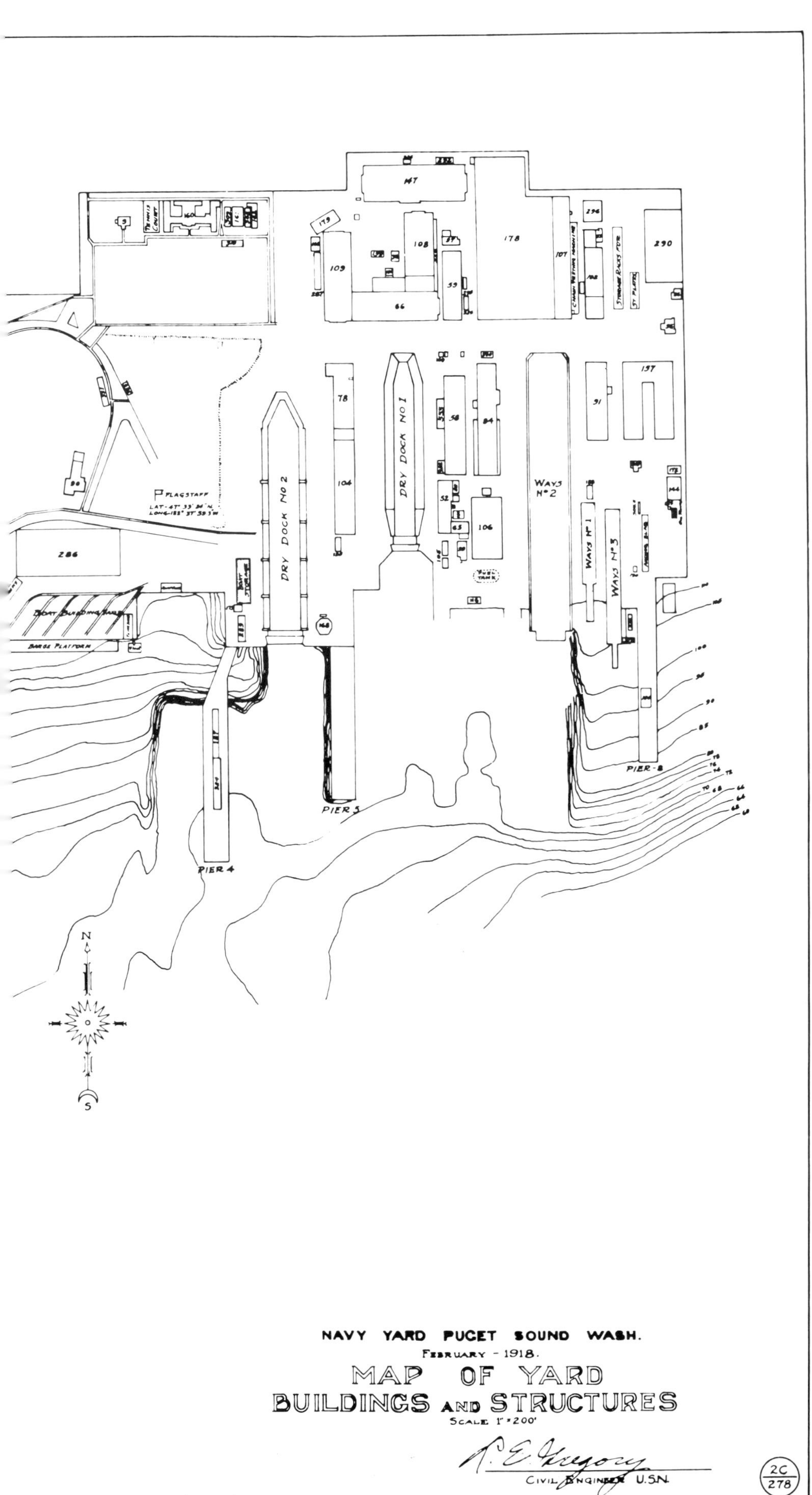

This 1918 map labels present Dry Dock 3 as "Ways No. 2." Not until 1921 did Dry Dock 3 receive its present name. Boat Building ways show on the waterfront south of the Stable (Bldg.90). The cemetery, which opened in August 1901, was crowded by the growth of the hospital in the northwest part of the Navy Yard. Barracks fill in the area west of the Marine Reservation.

Puget Sound Naval Shipyard

Men worked inside the caisson of the shipbuilding dry dock to install motors, pumps and other equipment for the operation of the caisson. In this July 1918 picture, Lieutenant Robert Morgan supervises riggers and shipfitters installing a steel plate of the caisson hull. John Morgan

The 1917-1918 Annual of the Bureau of Yards and Docks reported:

> *It was, of course, necessary with such a project that the caisson of the dock be built within the cofferdam in order that it should be in place when the cofferdam was removed, all of which required unusual and interesting construction. The work was prosecuted by the contractor along the line indicated, and the caisson was built by the yard force immediately above the sill in which it was to be placed, on blocking 8 feet above the dock floor. It was lowered by sand jacks for which no precedents could be found, the details of the same being worked out by the public works department of the navy yard.*[22]

Shortly before noon on December 16, 1919, Mrs. Gregory [23] opened the first valve to begin flooding the new dock. That afternoon, Mrs. Bisset, wife of Commander Guy Bisset of the Construction Corps, christened USS PYRO (AE 1) and Mrs. Henry Suzallo, wife of the President of the University of Washington, christened USS NITRO (AE 2).

In his 1919 report, Daniels referred to what is now Dry Dock 3 as "an innovation in American practice." At a cost of $600,000, the 927-foot long, 130-foot wide and 33-foot deep dock was the largest of its kind in the world. In addition to the two 460-foot-long ammunition ships, two target rafts had been built in the dock at the same time.

Before World War I, the Leadingman was responsible for the manner in which his trade accomplished the assigned task. There was essentially no prior design and the work of the draftsmen was to make an "as-built" record of the work the trades had already done.

The complexity of the work in the new ships demanded better work direction. This led to the establishment of a planning and estimating group whose function was to specify work packages for each trade. Thomas W. Winsor [24] moved from the electrical drafting room to take a leading position in this new group.

Winsor also saw the need to test the finished work. He and Charles C. Cowles [25] set up a laboratory in the loft of the electric shop to test repaired components, such as controllers and electrical motors, before they were re-installed in ships.[26]

A spin-off of this idea led the Yard into another high-tech field of work. World War I brought new gunfire control systems. These mechanical-electrical gun director systems were put into use for coast artillery and Navy shipboard guns. The Navy Yard's now established capability in

Children pose on top of the keelson, a steel fabrication bolted on top of a keel to strengthen it, at the ceremonial laying of the keel of Ammunition Ship No. 1, USS PYRO, in the shipbuilding dock, August 9, 1918. Puget Sound Naval Shipyard

On the day of the christening of Dry Dock 3 in December 1919, ammunition ships USS NITRO and USS PYRO were launched, as well as two small barges which were also built in the dock. That day signaled the entry of the Puget Sound Navy Yard into the world of shipbuilding. Photo was taken by H.E. Wale, an early photographer who used the clever business name, Prints of Wales. Puget Sound Naval Shipyard

Quarters U was built in early 1918 as the Radio Officer's home. In 1921, Quarters 377A and B were built for the Chief Petty Officers at the Radio Station. With further growth in that field, an enlisted mens barracks for radio operators opened in 1924.

Puget Sound Naval Shipyard

electrical testing techniques opened the door to work on these gunfire control systems. Making the remote indicating systems work was one of the tasks of the Director Shop, which was set up to install and repair devices produced by the Ford Instrument Company.

Technology involving sensitive shipboard instrumentation led the Navy to accept the Sperry Gyroscope as a primary navigation tool. William Wyler [27] came to work in the Yard as a gyrocompass engineer in 1918. With his help and expertise the Yard took on the new technology and established an outstanding reputation for high quality gyro compass work.

The Electric Shop helped develop the Navy's specifications for arc-welding motor generators. Winsor was involved with welding while the Navy was changing over from riveted hull construction to welded hulls. He developed a device for reducing the total power demand for the machines.

Because of the problems of matching metals with welding procedures, Winsor encouraged Fred Mills [28], to study metallurgy at Washington State University. Mills later became the Navy Yard's first metallurgist.

In addition to the training of recruits during the War, the Navy Yard was tasked with training radio men, and with the installation of radios on shipboard. Frank Marriott[29] headed the Radio Laboratory in Building 371, formerly the American Hotel on the land recently purchased at the east end of the Yard.

The Wireless Station, by then known as the Radio Station, was enlarged. During the war, Building 192 was constructed as the Radio Receiving Control building, and the Radio Officer and his family moved into a bungalow nearby.

In his 1921 report Secretary of the Navy Edwin Denby said:

The Navy's functions in time of peace are not confined solely to activities connected with its primary mission — preparation and readiness to act in defense of the Nation. It is engaged continuously in useful and humanitarian enterprises in all maritime waters. Its radio service practically girdles the world and is available to all ships everywhere, even on the Great Lakes. . .

During 1923, the Navy provided humanitarian assistance following earthquakes in Japan and Chile; at fires in Smyrna after the retreat of the Greek Army. U.S. Navy personnel also rescued people from a burning French transport that year.[30]

After World War I, the Navy announced that the Pacific fleet would be increased to the size of the Atlantic fleet. Because a larger industrial area would be needed to handle the increased workload, another feature of the 1916 Yard Development Plan [31] was put into effect. By cutting away some of the Yard's hillsides and using the excavated dirt to fill in the waterfront, the Navy Yard could increase the size of the industrial area.

In an April 1918 letter to the Navy Department, Coontz stated that the Navy Yard consisted of 228.7 acres, over one-half on high ground with about 78 acres of flat land in the industrial area and another 15.6 acres at the Training Center in the west end of the Yard.

He predicted that excavation would yield 34 acres of flat land and the fill at the waterfront another 49.5 acres for industrial growth. A dominant feature of the plan was to increase the expanse of land available for storage of supplies when the Yard would be a great fleet supply base.

The old Marine barracks was razed. The building that had been the Marine Officers' quarters and the four sets of

The former Marine Officers' quarters is shown here at its original site on the east side of Chester Avenue. Second Street (now Gregory Way) runs through the center of the picture, up and over the hill. At the left are radio station towers, Building 51 and a segment of the Navy Yard fence. At the right is Kitsap Inn. Kitsap County Historical Society

Apparently litterbugs were as prevalent 70 years ago as they are today. Upon approaching the Chester Street gate from inside the Yard, one read a sign, "It is forbidden to litter up the streets, walks and grounds with waste paper and other refuse." The Chester gate, which had replaced the Warren Avenue gate, closed in December 1921, when the State Avenue gate opened. Puget Sound Naval Shipyard

Seen from the officers' quarters, this view shows some results of the earth-moving job accomplished by contractor Porter Brothers in 1919. In front of the quarters, the ground was lowered 70 feet, and 50 feet at the Marine reservation. The old stable, built in 1902 and enlarged through the years, stands right center. Behind the stable, Buildings 78 and 104 show the connecting section built during World War I.

Puget Sound Naval Shipyard

This November 1920 photo, looking westward toward the officers' quarters, shows excavation and grading work almost completed and gives a graphic view of the impressive scope of work involved. Quarters S, to the right of the flag pole and partially hidden by trees, had been moved to that location from the area shown already graded in the foreground.

Puget Sound Naval Shipyard

The most critical part of the Yard Development Plan depended upon obtaining enough flat land for the installation of buildings and other facilities needed to up-grade the Navy Yard's capabilities. By 1922, the filled-in waterfront west of Pier 4 was occupied by a few new buildings. The waterfront between the gravel bunker pier, visible here with crane and a cloud of steam; and the fueling pier at the left edge of the picture, is the western part of the present Shipyard waterfront, ending with Dry Dock 6. From the fueling pier to the coaling wharf, upper center, is the site of the present Supply Center.

Larry Jacobson

quarters west of the ravine were moved to the ends of the Navy Officers' quarters.[32]

Porter Brothers of Seattle regraded the land, their huge steam shovels biting out sand and gravel to be transported to the waterfront for fill. Lost were the lovely trees and the golf course Burwell had established on the gentle slope below the quarters. Four steam shovels, many small engines and several hundred men moved more earth than had been moved up to that time in Seattle's regrade.[33]

Early plans called for heather to be planted on the steep slope created by the excavation, but the rapid erosion of the hillside led to the choice of more easily established Scotch broom. Later, the slope's expanse of golden blooms became a tourist attraction.

An 1,100-foot reinforced-concrete sea wall was built along the new waterfront. The seawall provided berthing for shallow draft vessels, in addition to its primary purpose of preventing erosion of the fill from the excavation.

All Navy Yards were put on the same system of management in 1921: Commandants were given full authority to run the Yards and were made solely responsible for results. Several good features of the previous system were kept, including the retention of a manager in charge of the industrial activities of the Yard and responsible to the commandant alone.

The first International Conference on Limitation of Naval Armaments met in Washington, D.C., in November 1921. After lengthy bargaining, it was agreed the countries involved would scrap many of their ships, whether already built or still under construction, and would limit future building. The Washington Naval Treaty signed February 6, 1922, established a 5:5:3:1 3/4:1 3/4 ratio in battleship and aircraft carrier tonnage among United States, Great Britain, Japan, France and Italy.[34]

No agreement was reached on construction limitations on other types of vessels. This conference and at least six others failed to impose lasting limitations on naval construction internationally. The U.S. adhered to the limitations and upgraded its naval efficiency by disposing of overage ships and obsolete equipment.

By April 1921, the Yard had completed all the work that had been authorized during the war. During the next seven years the Yard built only two ships: the repair ship, USS MEDUSA commissioned in 1924 and the submarine tender USS HOLLAND in 1926.

Repair Ship USS MEDUSA (AR 1) glides into Sinclair Inlet after her launching April 16, 1923. Leaving Dry Dock 3, she moved to her fitting-out pier where she was later commissioned. Puget Sound Naval Shipyard

During these years, the increased number of ships in the Pacific Fleet provided repair work for a much reduced work force. Denby's 1923 report praised the Yard:

> *On the transfer of the fleet to the west coast it has become necessary to do the work of regunning of ships at the Navy Yard, Puget Sound and to do the necessary minor repairs and testing of torpedoes belonging to vessels of the fleet at the Naval Torpedo Station, Keyport, Washington. During the year, three capital ships have had their main batteries renewed and their old batteries removed and returned to the Naval Gun Factory for relining. In the case of the TEXAS the Navy Yard, Puget Sound made a very notable record by removing the ten 14-inch guns belonging to that ship and replacing them with a new battery in the time of 47½ hours.*

The largest single contract let by the Navy during fiscal year 1922 was for the rebuilding of what is now Pier 5 at a cost of $637,724. An old wooden pier was removed and replaced by a new 80-foot wide and 1,200-foot long reinforced concrete pier. The pier was completed in 1922 and serviced by a 50-ton traveling crane.[35] The Navy Yard's first permanent pier (Pier 4), which had been built following the construction of Dry Dock 2, was lengthened.

Six young men had entered the Navy Yard in 1901 as apprentices. One, Arthur Holden,[36] was hired as a "minor-under-instruction" and trained as a draftsman; no academic instruction was planned for the others.

At the request of the apprentices, a school was begun in 1914 with a few desks and an old blackboard on the second floor of the old Joiner Shop (Bldg. 91). Naval officers were the instructors. From 1915 to 1917, school was held in the Disciplinary Barracks (former Marine Barracks); in 1917, in Building 78. Classes in 1919 began in "the tin castle", formerly the buffer shop, but rain on its corrugated metal roof was too distracting. The school then moved to the old Labor Board Building (Bldg. 56) near the main gate and in 1920 moved to Building 50, where the school remained until 1954.

The school shared Building 50 with the Orr and Fisher families. Jesse M. Orr became the Yard's fire chief in 1921, and Fisher was the Commandant's chauffeur. It was a wonderful time for young Richard Fisher and Howard Orr.[37] Howard experienced the thrills of riding to fires with his Dad.

The Yard had purchased its first fire engine in 1919; the American-LaFrance pumper was housed in the Stable

By 1927 the Yard was ready to fight oil fires, as evidenced by this fire truck identified as N.Y.P.S. Chemical No. 1. It was equipped with two sizes of portable fire extinguishers as well as bell and siren. Puget Sound Naval Shipyard

(Bldg. 90), and one employee was detailed to work the pumper. Firefighting crews in several shops raced to fires pulling small hose carts. All workers assigned to firefighting were chosen from employees living near the Yard as they were also on call after working hours. Orr, a crane operator who had been fire marshal at Todd Shipyard during World War I, was selected as fire chief, when the decision was made to up-grade the system.

Children growing up in the Yard received a practical education from their exposure to many facets of life. They knew the lure of the great ships, understood the language of the Navy and the Yard, and were familiar with one or more trades. Like many others whose parents and grandparents worked in the Yard, Orr and Fisher became employees. Howard Orr retired as a crane operator in the Boat and Joiner Shop. Fisher, who learned his multiplication tables at a big desk in the old stables where the car was kept, worked in the Transportation Shop.

Other children who grew up in the Yard had careers in the military service.[38] Admiral Noel Gayler remembered setting traps for mountain beaver in the new golf course set up in the ravine after the removal of the front hill.[39] His father was Public Works officer for three tours of duty between 1920 and 1940. His mother, Ann Gayler, in her memoirs, remembered:

> *"Little children living in the white houses on the hill dressed early to march behind the Marines and the Navy Band to raising of the colors. They were allowed to march on the return as far as the brink of the hill where it dipped down to the Marine Parade Ground, and then home to breakfast and school. In the long summer twilights, they scurried home from play at the sound of the sunset gun."* [40]

The Bremerton Chamber of Commerce donated a 1.5-acre campsite with 350 feet of shoreline on Kitsap Lake to Rear Admiral Josiah McKean[41] in June 1921, for the use of "the men of the Navy." The BREMERTON NEWS reported that it was put to immediate use by a camping party from USS ARKANSAS, who enjoyed the lake site and the trails leading back into the forest and waterfalls.

The Chamber deeded an adjacent 12.5 acres of land to the Commandant of the Thirteenth Naval District on January 18, 1924, for one dollar. On March 24 of that year, an additional 8.08 acres of land was purchased from H. McChesney for $1,010.[42]

During the peace years of the 1920s, the non-military areas of the Navy Yard were developed to meet the needs of personnel in the increased number of ships in the Pacific. The need for indoor recreation for the patients and staff at

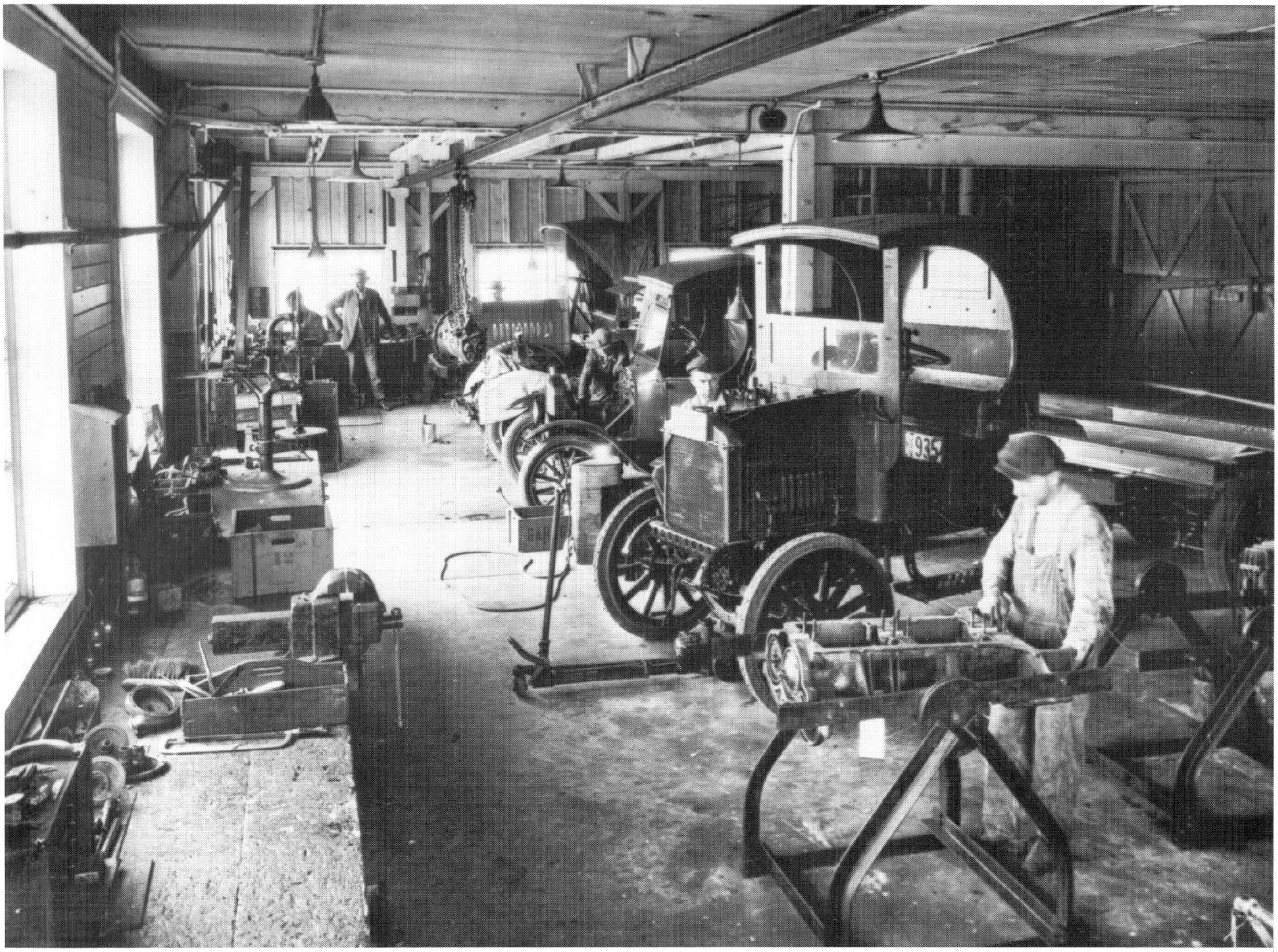

Inside the Navy Yard garage, mechanics work on trucks and cars in October 1927. The man at the right rebuilds an engine. Although several lighting fixtures are apparent, most of the light comes through the windows. Today that lighting would be considered inadequate. The low ceiling contrasts to one in a modern commercial garage.

Puget Sound Naval Shipyard

Building 400 has been used for the Yard's swimming pool since it was built in 1920. Other than adding skylights and boarding over the window's lower halves, the building has had few changes over the years. The photo shows the pool inside the 90x34-foot building. *Centennial Book Committee*

the hospital was met in December 1920 when the Red Cross provided a "Convalescent Home" (Bldg. 379). This was built on Dewey Street near the present Naval Avenue gate. It was completely furnished and included bowling alleys, canteen, lockers and a large assembly room with stage and moving picture equipment.

A swimming pool (Bldg. 400) was erected in 1920 at the west end of the Yard.[43] That same year the Navy Yard Cemetery, situated between the Red Cross Building and the fence, was removed. Many of the bodies were claimed by relatives or friends; 65 others were transported via USS NITRO to the Presidio National Cemetery in San Francisco for reinterment.[44]

An east wing was built on the hospital in 1922, completing the originally planned facade. The nurses moved into quarters (Bldg. 376) south of the Red Cross Building in 1921, and two years later a two-story addition was completed. Quarters were built for doctors in the mid-1920s.

Alonzo Johnson,[45] who served as a corpsman at the hospital in the early 1920s, remembered the Chinese culinary staff that worked in the Commissary, as the Food Service Department was called.

The Navy Yard Cemetery, shown here next to the Red Cross Building, received its first deceased sailor, Seaman Rankin, in August 1901. Almost 100 bodies were buried here before they were exhumed to make room for hospital expansion. The most distinguished person interred was Medal of Honor recipient Axel Westermark, a hero of the Boxer rebellion.

Puget Sound Naval Shipyard

Under construction in January 1924, is a second hospital ward, built by Yard labor, to the north of the existing west ward. The Nurses' quarters stand left in the background with the Red Cross Building to the right. *Kitsap County Historical Society*

Randina Olsen, first woman to retire from Puget Sound Navy Yard, stands with friends in front of her home at Sixth Street and Hewitt Avenue, Bremerton, in April 1925. Front row left to right are Lyle and Chester Robards, Ruth Crane (Cloaherty), Bobby Lent, Dina's granddaughter Jean Henry (Jensen), Betty Miller and Bobby Dewey. Second row: Mr. and Mrs. Rice, fellow worker Bertha Gillespie, Dina Olsen, Dina's Leadingman Mr. Miller and Mrs. Miller, an unidentified woman and Mr. and Mrs. Lawrence Brook. Third row includes Frank Iverson, a Public Works Department truck driver, Mrs. Charlie Horn, Mr. and Mrs. Chris Rasmussen and Dina's daughter Mable Henry behind Mrs. Miller. Navy Yard band members, who had escorted the group out of the Yard, complete the picture. Jean Jensen

These Chinese who were all of Cantonese origin, were not under the U.S. Civil Service and could be hired or fired by the Hospital Commanding Officer. An administrative head or supervisor, also foreign born Chinese, was overseer of nearly all aspects of their activities. Few were English speaking and little English was heard around that department. Although foreign born and self-taught, Lee Yok had an uncanny command of English, including Navy expressions and slang.

He ran his crew with a strict discipline and at times forced "siestas" on them when he wanted to insure strict physical fitness when he knew that extra exertions were going to be required especially around Christmas or holidays. To a degree this place acted as a school for these Chinese because after a few years when they had learned English and American ways, they would resign and go out and open a laundry or Chinese restaurant. When this happened another non-English speaking "cousin" would arrive for the vacant job.

The Civil Service Retirement Act of 1920 fixed the mandatory retirement age of federal employees at 65. No records have been found indicating who the first person was to retire from the Puget Sound Navy Yard, but in a May 23, 1947, Salute interview, 89-year-old Ole Olsen said he retired in 1923. Olsen entered the Navy Yard in 1902 as its first sailmaker, going to work in a small shack where the Sheet-

Before the Navy Yard's use of paychecks, workers were paid in cash by the Paymaster, who traveled on foot from shop to shop, or later from a pay wagon. Here the vehicle is decorated for a parade during the 1920s. Alice Lawson

metal Shop now stands. Later, as a leadingman with six other sailmakers, he set up shop in a building where Dry Dock 5 is now.

By that time, there were women working in the Sail Loft. Dina Olsen (no relation) began there in 1905. She retired April 10, 1925, and the BREMERTON NEWS AND SEARCHLIGHT declared her to be the first woman to retire from any United States Navy Yard.[46]

Until the late 1930s, Navy Yard employees were paid in cash, a pay wagon appearing at various sites on a set schedule. A 1925 circular from the Navy Department encouraged the practice of using $2 bills but left the use of silver dollars to the discretion of the Commandant. In January 1928, pay days were set for each Friday instead of the 8th, 15th, 23nd [47] or last day of the month as previously done. A foreman, quarterman or leading man was required to be present to identify employees, making sure they were in their proper places in line and had signed their vouchers in ink.

In spite of efforts to keep a larger force busy, employment was down to less than 2,500 by 1927. Congress had authorized the construction of eight 10,000-ton light cruisers to bring the U.S. up to the limit set by the Arms Limitation Conference. Military and civilians supporters of the Navy Yard directed their efforts to having construction of one of the cruisers assigned to the Navy Yard at Puget Sound.

Congress appropriated the money to build the last three cruisers in 1927. That Spring the good news came. The Yard would build No. 28, under the first contract the Yard had ever received for a surface combatant ship.

It would be the first of many.

Symbol of the Navy Yard's new mission, USS LOUISVILLE, still fully dressed after her launching ceremony, is shown being warped into her fitting-out berth where her construction would be completed. The date was September 1, 1930.

Puget Sound Naval Shipyard

Chapter Four
Through the Depression
1928-1941

When the Puget Sound Navy Yard began work on Cruiser No. 28 (later named USS LOUISVILLE) on July 4, 1928, most of the United States was enjoying prosperity. Black Tuesday and the ensuing Great Depression were more than a year away. However, people in the Puget Sound Navy Yard area welcomed LOUISVILLE's construction as an end to lean times.

USS LOUISVILLE[1] was one of the 10,000-ton cruisers permitted under treaty limits and authorized by Congress in 1924. President Calvin Coolidge had delayed the funding in a vain attempt to entice Japan into extending the Naval Building Moratorium[2].

PSNY's submitted estimate of $8,599,250[3] fell far below Congress' $11,100,000 per ship statutory limit and $2,000,000 under those of private contractors' bids for sister ships. It was a daring gamble. The Yard had never built a ship of this size and complexity; however, Navy Yard Commandant Rear Admiral Henry J. Ziegemeier had great confidence in the Yard's work force. Securing this much needed contract brought an increase in the work force from its post-war low of fewer than 2,500 employees in 1927 to more than 4,000 in 1929.

The keel was laid July 4, 1928, with:

> *. . . the crackle of firecrackers, and eerie whistle of pinwheels, the heart thumping sound of 'Under the Double Eagle' . . . That was America, Circa 1928. A vintage year in national expression of patriotism, in worker pride of accomplishment of the impossible when the chips were down.*[4]

USS LOUISVILLE was launched with all due ceremony on September 1, 1930. Critics had considered Puget Sound Navy Yard unable to do the job at the bid price because of lack of experience. Confidence and co-operation between military and civilian personnel proved that the Yard could do it, and at less than estimated cost.[5]

Much credit for the successful construction of LOUISVILLE was given to Ziegemeier's leadership. Unfortunately, he did not live to enjoy the triumph; he suffered a heart attack and died October 15, 1930.[6]

As a memorial to the deceased commandant, the newly completed yard theater was named for him. Ziegemeier Theater, seating 298, showed the new "talkies" and was used for official ceremonial activities such as Apprentice School graduation exercises. The building, with its recessed romanesque arches, fell to a wrecking ball in the 1980s to make space for a parking lot.

The Yard's successful performance on LOUISVILLE played a large part in PSNY receiving a contract to construct USS ASTORIA (CA 34). Once again the Yard's bid was $2,000,000 under the limit set by Congress.

The keel for ASTORIA was laid in Dry Dock 3 immediately following the launching of LOUISVILLE and the draining of the dry dock. ASTORIA was launched on December 16, 1933, sponsored by Miss Leila McKay, a great-granddaughter of one of J. J. Astor's partners in the fur trade.

While work began on LOUISVILLE, military and civilians welcomed the arrival of two ships, USS LEXINGTON (CV 2) and USS SARATOGA (CV 3), which signalled the Navy Yard's entry into a new field: aircraft carrier repair and conversion. The ships had been built on the

Members of the reception committee for USS LOUISVILLE's keel-laying pose alongside the cruiser's keel. Standing are R. Canfield and W. W. Brock of the Technologist Association, H. B. Richards and J. H. Warren of the Master Mechanics and Foremen Association, R. H. Lindahl of the Federal Employees Union and O. C. Jackson of the Apprentice Association. Seated are H. Stover of the Supervisors Association, Hull Production Superintendent Lieutenant Commander C. F. Osborn and H. Chandler of the Apprentice Association.

Puget Sound Naval Shipyard

USS LOUISVILLE's keel and the multitude of guests participating in the keel-laying ceremony show graphically the relationship of the shallow depth of the dry dock to its length and width. On LOUISVILLE's launching date, USS ASTORIA's keel would be laid in what is here the guest seating area.

Larry J. Jacobson

Rear Admiral Henry J. Ziegemeier, wife Jewel and five-year old daughter Rosemary pose on the hill above the industrial area. Not long after this picture was taken, tragedy struck the family when Ziegemeier suffered a heart attack and died.

Centennial Book Committee

The launching party and invited guests assemble for USS LOUISVILLE's launching on September 1, 1930. The crane to the right is hoisting the keel for USS ASTORIA, to be laid that same day.

Kitsap County Historical Society

hulls of two battle cruisers whose completion had been halted by treaty limitations.

The two carriers, commissioned late in 1927, came to PSNY in 1928 for replacement of defective turbine blading. They returned in the fall of the following year for additional work. USS LANGLEY (CV 1), the converted collier JUPITER, joined them in the Yard that November.[7]

Although there had been plans for a naval air station south of Gorst, in 1922 the Navy leased the Sand Point acreage that King County had purchased for an airfield. In 1926, King County deeded 411 acres to the Navy Department for a Naval Reserve Air Station.[8]

The Navy Yard played a minor role in the Navy's experiments with lighter-than-air craft. Riggers were sent to Camp Lewis near Tacoma in 1924 to set a mooring mast for the dirigible USS SHENANDOAH. She arrived in the Puget Sound area after a successful voyage from Lakehurst, New Jersey, by way of San Diego.

Amphibious aircraft training constituted 40 percent of Naval pilots' training in the mid 1920s.[9] To accommodate the flying boats at the Navy Yard, a seaplane ramp was built in the late 1920s, south of Building 457.[10]

On their first availabilities in the Navy Yard, LEXINGTON and SARATOGA were too large for any of the Yard's three docks; work was done at pier side. In order to accommodate these large ships, plans were made to lengthen Dry Dock 2. Getting appropriations to do the job was another matter.

In a March 1929 letter to the Bremerton Chamber of Commerce, Representative John F. Miller stated that Rear Admiral Luther Gregory, then Chief of the Bureau of Yards and Docks, called the dry dock extension the most important item in the Navy's budget as far as the welfare of the fleet was concerned. However, it was only after long and heated debate that Congress appropriated the money in 1930. That year a notch cut in the head of Dry Dock 2 extended it to a length of 867 feet.

Electric drive ships, such as SARATOGA and LEXINGTON, provided much of the work of the Navy Yard at this time.[11] SARA, as SARATOGA was affectionately known, was a frequent visitor through all her years. Secretary of the

All three of the Navy's aircraft carriers tied up at Puget Sound Navy Yard piers in November 1929. Looking south we see USS LEXINGTON (CV 2), USS SARATOGA (CV 3) and USS LANGLEY (CV 1), a converted collier. Larry J. Jacobson

This November 1931 view, looking northwest, shows two float planes resting on their pontoons at the seaplane landing southeast of the State Avenue Gate. Houses on the west side of State Avenue are visible behind the flag pole.

Puget Sound Naval Shipyard

The new section of Dry Dock 2 added 40 feet to its north end, bringing the total length to 867 feet, and allowing it to dock the largest Navy ships of that time, the aircraft carriers. Photo taken October 21, 1930.

Puget Sound Naval Shipyard

The Enlisted Mens' Club opened in Building 422, west of the movie theater, on June 30, 1930. An addition in 1942 created room for a Ships Service (Navy Exchange) and Beauty Shop; the latter operated there from 1943 until the building was converted to a Waves' dormitory; in November 1960 the Chief Petty Officers took over the structure. At various times sections of the building were used as a day care center, Navy Relief offices and Thrift Shop, and Sunday School. The building was demolished in 1987.

Puget Sound Naval Shipyard

An early 1930s aerial view reveals the Yard, lush forests of South Kitsap County, the glory of Mount Rainier and calm waters of Sinclair Inlet. On the left are the Hammerhead Crane and several ships. The Marine Barracks is clearly visible in the lower center, as are the sand traps and greens of the golf course.

Centennial Book Committee

On September 25, 1930, Crane 27 lifted a turbine from SS EMPRESS OF RUSSIA for removal to the Machine Shop (Bldg. 66) for repairs. The crane built by McMylar Interstate Company in January 1923, served the Yard until scrapped in 1952. According to the February 8, 1952 SALUTE, three of her former operators, Harry Adams, Lester E. Holland and Kenneth A. Cramer gathered for a last farewell to what had been the Navy Yard's most powerful crane.

Puget Sound Naval Shipyard

Navy Charles F. Adams' report to President Herbert Hoover in 1931 made note of research by the Metallurgical Laboratory and Welding Shop at Puget Sound Navy Yard. Pitted areas of SARATOGA's propellors were repaired by filling the eroded areas with manganese bronze, using an arc-welding technique.

Tacoma considered LEXINGTON a heroine. Drought during the fall of 1929 and an unusually early cold spell threatened to prevent a sufficient flow of water over the dams at Lake Cushman and the Nisqually River, sources of Tacoma's hydro-electric power. City officials requested the loan of LEXINGTON to generate power for the city.

On December 15, 1929, the huge ship tied up at Baker Dock on Tacoma's waterfront. After electrical connections had been made, LEXINGTON began transmitting 20,000 kilowatts of power for 12 hours a day. This allowed water to build up behind the dams. Mission accomplished, LEXINGTON left for maneuvers on January 16, 1930.[12]

The following year LEXINGTON sped to Managua, Nicaragua, where her planes delivered medical personnel and supplies for disaster relief after an earthquake and fire had practically destroyed that city. Another ship on a mission of mercy that year was the newly commissioned USS LOUISVILLE, whose personnel rescued passengers and crew of a ship grounded off Point Arguello, California.

The Yard's facilities for docking, machining and handling large equipment continued to induce other countries to send vessels to PSNY for repairs. The Canadian Pacific Railway Company sent the SS EMPRESS OF RUSSIA to the Navy Yard in August 1930. She was one of the early turbine-driven ocean liners and required extensive repairs, including corrections in design and balancing of main engine turbine rotors. Roger Pacquette [13] recalled the day the EMPRESS OF RUSSIA welcomed visitors:

> *. . . on a Saturday afternoon people swarmed aboard the Empress and looked into every nook and cranny. The crew, with the exception of the officers, were Chinese, who were not allowed ashore. They were living, eating and gambling just as they would in their native country. They* (had) *supplemented their rice diet by fishing out of port holes while the drydock was being emptied. In those years the emptying of the drydock at the right season, could leave the bottom of the dock completely covered with herring and smelt, and now and then salmon.*

A very different ship came to the Yard in 1933. Early in the century when the Navy announced its decision to scrap all wooden ships, there had been a nationwide outcry to save CONSTITUTION, better known as "Old Ironsides".[14] Later, plans were made for a complete restoration of the ship. On April 17, 1929, Ziegemeier sent four car loads of Douglas fir for the ship's masts and spars. This wood was the gift of the West Coast Lumbermen's Association. $300,000 was appropriated for the restoration in 1930, and additional funds were raised by public donations. School children throughout the country sent their pennies.

"Old Ironsides" was recommissioned in Boston on July 1, 1931, and began a cruise to 90 ports on the Atlantic, Gulf and Pacific coasts, accompanied by USS GREBE (AM 43).[15] They arrived at the Puget Sound Navy Yard on July 1, 1933. Bremerton[16] Mayor J. A. McGillivray headed the reception committee which welcomed the ship to the city. A total of 28,092 persons toured CONSTITUTION from July 2 through 6. That brought the total number of people, who had visited the ship since her re-commissioning, to 3,905,781.[17]

As a consequence of lessons learned from a disastrous

Independence Day 1933. "Old Ironsides", USS CONSTITUTION, is moored starboard side to the Receiving Ship pier, which she shares with her escort USS GREBE (AM 43). A ferry approaches the southern shore of Sinclair Inlet in this Harry Ward photo. Kitsap County Historical Society

A Nordic princess, skivvy-clad skeleton and hairy hula dancer brought hilarity to a basketball game between Masters and Supervisors in March 1933. Here the Masters pose as champions of the Navy Yard: Master Molder John L. Sendner, Master Painter Henry C. Jaixon, Master Machinist W. E. Abbot, Master Shipwright and Dock Master George M. Trahey, Master Boatbuilder and Joiner Charles V. McLaughlin, Master Pipefitter Julius Hauschel, Master Machinist F. A. Smith, Master Shipfitter Herman Petersen and Master Boilermaker George W. Penketh. Puget Sound Naval Shipyard

1931 reconstruction of Dry Dock 1, in which the wooden side walls were replaced with concrete, was both time consuming and costly. Reconstruction costs exceeded half the price of the original dry dock. Puget Sound Naval Shipyard

fire at the Naval Ammunition Depot at Lake Denmark, New Jersey in 1926, the Navy required greater separation between individual ammunition storage areas. Additional land was acquired for the Torpedo Station at Keyport and at the Ammunition Depot on Ostrich Bay in 1929 and 1930, making this wider separation possible. The new bunkers were officially described as being "of the latest design with curved roofs to shunt off falling objects which might otherwise cause explosion by shock." [18]. Although emptied long ago, some "igloos" are still visible at NAD Park and Jackson Park Housing.

The memorable Good Will Apprentice Basketball series began in 1930. Former PSNY apprentice Arthur G. "Smitty" Smith,[19] who worked at Mare Island, wrote to Walter Biles, a former classmate, outlining the idea of a series between the two West Coast Navy Yard Apprentice Basketball teams. Biles interested the Apprentice Supervisor and Master Mechanics in the project; the Masters issued a challenge to their counterparts at Mare Island to field an apprentice basketball team and to begin a three-game series in Bremerton in February 1930.

Mare Island won the series that year, and the next at Vallejo. Puget Sound won in 1932. The trophy went back and forth between the two Navy Yards until 1954. That year Puget Sound began a stretch of three successive wins that gave it permanent possession of the trophy, still on display at the Apprentice School. The friendships developed during the series provided a great deal of good will that benefited the Navy through the close co-operation between the former apprentices who became leaders at the two Navy Yards.

A major requirement of industrial installations is to have enough work to maintain a stable workforce. In a special message to Congress on December 4, 1930, President Herbert Hoover requested appropriations for a group of public works projects which would fulfill a definite naval need, while improving the nation's employment situation.

Much of the responsibility for this work at the Puget Sound Navy Yard fell on Lieutenant Commander Ben Moreel,[20] who was Public Works Officer for the Yard and Thirteenth Naval District from June 1930 to May 1932. His ability had already been recognized; as a lieutenant commander he held a postion normally assigned to a captain.

One of the projects during Moreel's tour was a $100,000 Transportation Equipment Building (Bldg. 427), which later housed the Electric Shop. Another major project completed in 1931 was the removal of the wooden walls and complete reconstruction in concrete of Dry Dock 1, during which it was shortened to 639 feet.[21] At the Central Power plant, yard workmen installed a new 6,000 kilowatt turbo-generator and constructed a reinforced concrete smoke stack.

Beginning in February 1931, Hoover's program paid for approximately 5 percent of all work done at Navy Yards and Stations throughout the country. In April of that year, those funds paid 30 per cent of the work force at PSNS.

This 1932 panorama shows USS LEXINGTON moored at Pier 6, Crane-ship KEARSARGE, battleships, the cranes

Shortly after taking office in 1933, President Franklin Roosevelt announced the inauguration of the National Industrial Recovery Act (NIRA) as the "machinery necessary for a great cooperative movement throughout industry in order to relieve unemployment." Under its auspices, the local Navy Yard communities received a real depression breaker: contracts for construction of destroyers.

The Yard received authorization October 1, 1932 to build the 1500-ton destroyer, USS WORDEN (DD 352). She belonged to the FARRAGUT class, the first destroyers to be

around Dry Dock 3, smoke-stacks on Buildings 91 and 106 and Supply Department Building 157 in the foreground.
Centennial Book Committee

built since World War I. Katrina Loomis Halligan, wife of the Commandant, was sponsor when WORDEN was christened on October 27, 1934.

Japan had invaded Manchuria in 1931; in 1934 the Nazi party came into power in Germany. As the expansionary aims of these two countries became apparent, the stability of the world was threatened. Congress reacted by passing the Vinson-Trammel Act of 1934, thereby permitting the Navy to build up the fleet to the limits allowed by the Washington and London Arms Limitation Treaties of 1922 and 1930.

This section of the panorama shows the north end of Building 157, shops extending along Farragut Avenue, and Building

Authorization to build five destroyers followed: USS CUSHING (DD 376) and USS PERKINS (DD 377) were launched on December 31, 1935, USS PATTERSON (DD 392) and USS JARVIS (DD 393) on May 6, 1937 and USS WILSON (DD 408) on April 12, 1939.

The Yard's work force rose steadily. New skills were needed; many of the men who were recruited had learned their craft in other lands. Hal Halvorson [22] knew many of the men who:

" . . . *came from the Banks of the Clyde. They were*

290 on the right. The stores in the foreground are on land in Bremerton purchased during World War II.

Centennial Book Committee

very competent, took great pride in their work and had a strong desire to share their knowledge and expertise. One of the most used expressions was "come on laddie, gi your tools and lets go to work."

Under the NIRA program, $1,908,150 was appropriated for public works projects in the Thirteenth Naval District, most of which was spent at the Puget Sound Navy Yard. The first major result of this appropriation was the hammerhead crane, built at the south end of Pier 6 in April 1933 by the Dravo Company of Pittsburgh, Pennsylvania. The crane, symbol of the present Puget Sound Naval Shipyard, was built with a rated lifting capacity of 250 tons and was tested at 350 tons.[23]

The Hammerhead Crane, rated at 250-ton safe working load is shown here lifting a 350 ton test load consisting of four 16-inch guns, four eight-inch guns and a test weight suspended from chain falls. The men on the ends of the gun barrels are participating in the testing procedure. Puget Sound Naval Shipyard

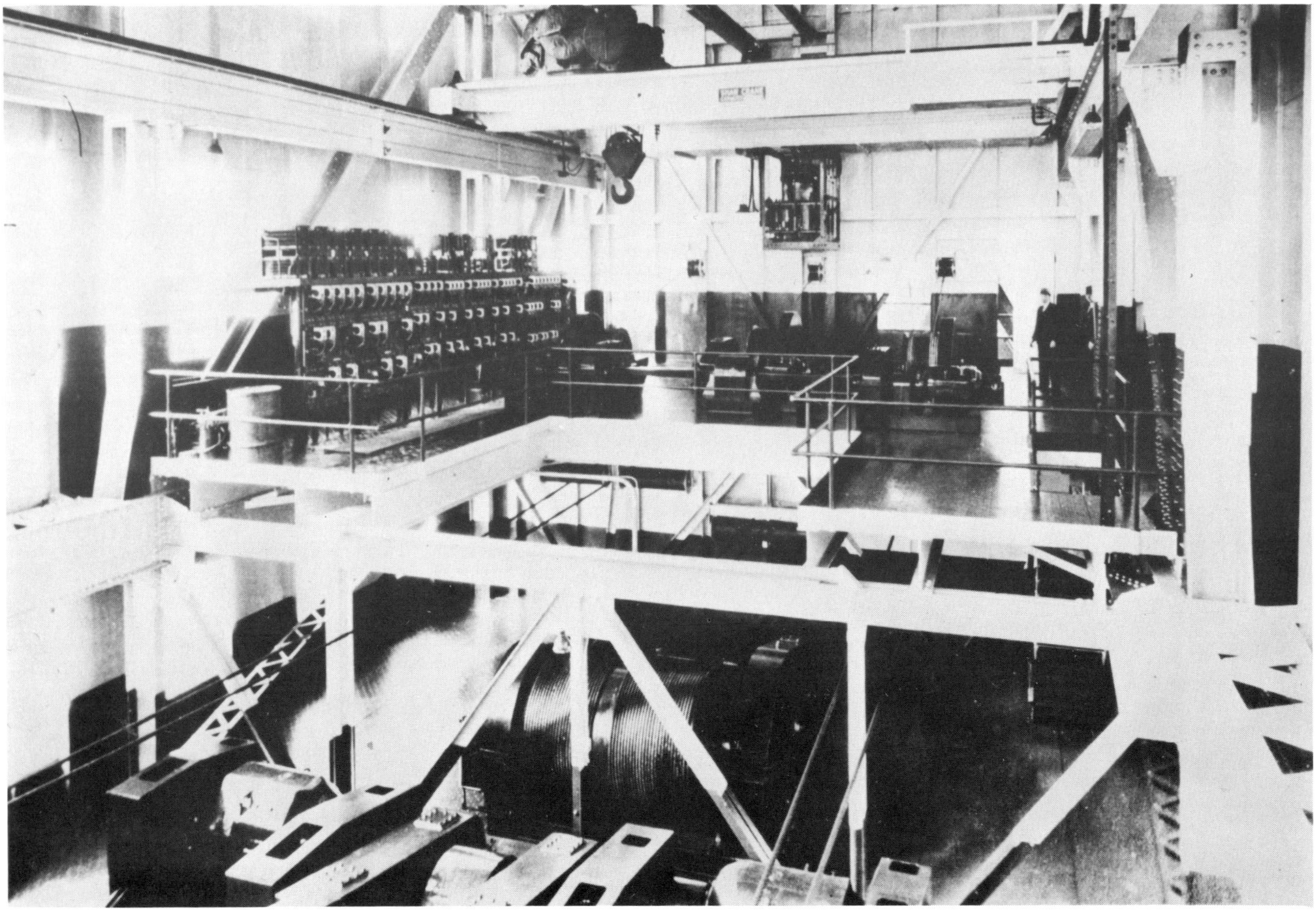

The machinery house, high above Pier 6, contains equipment for operating the Hammerhead Crane. The three-story room is reached by elevator. A sub-station at the crane's base supplies power to the lifting motors. Puget Sound Naval Shipyard

Building 431, containing the Machine and Electric Shops, and USS OKLAHOMA are reflected in Sinclair Inlet. The 800 by 250-foot structure has been augmented in length, width and height over the years. It now measures 979 by 349-feet, with a seven-story tower (not shown in this early photo) for periscope repair work. Puget Sound Naval Shipyard

The major portion of the appropriation, $1,350,000, financed a new Machine and Electric Shop (Bldg. 431). The new building, 800-feet long, 250-feet wide and 94½ feet high in the center bay, dwarfed the old Machine Shop, Building 66, north of Dry Dock 1.

On its completion in June 1935, Building 431 was the largest machine shop west of the Mississippi River and considered the finest of its kind in the nation. The 1938 Navy Day Annual[24] reported the building was constructed of heavy concrete and steel with floor space totaling five acres. The walls were constructed almost entirely of special glass set in a new type of steel sash that allowed a maximum amount of natural light. In an emergency, the building could handle 1,000 workers per shift.

The NIRA fund also paid for the removal of the coaling plant and its pier. By 1930 the Navy had become almost exclusively oil burning, and all naval coal storage stations on the Pacific coast were inactive. At Puget Sound, the excess coal was converted into briquettes for local domestic use.

As Roosevelt's New Deal progressed, new agencies were formed and acronyms multiplied. The NIRA was joined by the WPA (Works Progress Administration), CCC (Civilian Construction Corps) and others. Under the CCC program, young men learned skills as they razed, excavated and built structures throughout the country. CCC projects for the Navy in the Northwest were at the Ammunition Depot on Ostrich Bay. The men built two homes, a guard station and a stable.[25]

Late in August 1935, approximately 1,700 men on relief (Federal financial assistance) started to work on Navy projects. Although the men were carried on the WPA rolls, they were supervised by Navy Yard foremen. Improving the electrical distribution system formed a major part of their work; roads, railroad tracks and walks were reconditioned, after the wiring was placed underground. They performed general cleanup work and re-painted buildings and fences. At this time, Building 438 was erected for the custody and issuing of tools.[26]

PSNY also manufactured various articles; production of valves and small boats kept an average of 275 men a day busy. The Yard constructed several diving boats in 1937 and in 1938 built 30-foot motor launches for the Coast and Geodetic Survey. The Yard built an average of 40 to 50 new craft a year.

This aerial view looking northeast concentrates on the yard's waterfront. USS SARATOGA(CV 3) is moored at Pier 6. Also present are USS WEST VIRGINIA (BB 48) in Dry Dock 2, USS ARIZONA (BB 39) at Pier 5, USS NECHES (AO 5) in Dry Dock 1 and USS JASON (AC 12) and KEARSARGE (AB 1) at Pier 4. This Prints of Wales' photo accompanied an April 1932 Bremerton newspaper article, which reported Bremerton and the Navy Yard looked forward to a bright future because of recent legislation. Centennial Book Committee

The hospital acquired a new garage, utility building, and a three-story brick and concrete dormitory (Bldg. 443) designed for 101 hospital corpsmen. The dormitory was located at the site of the old cemetery.[27]

By 1935 the old World War I barracks in the west end of the Yard had been razed and the grounds cleared and graded. Building 433, a four-story yellow brick structure, designed for handling all Receiving Station activities, was completed near the Charleston Gate.

Rear Admiral Thomas T. Craven served as Commandant of the Yard from July 1935 to July 1937. The closing days of his tour saw the culmination of one of his favorite projects. Craven was well known for his concern for the welfare of enlisted men and he wanted them to have a covered recreation area for use in winter and inclement weather.

The old Navy Yard Hotel, between Burwell and Fourth Streets, seemed appropriate for his purposes. The four-story hotel had never been a financial success because of the drop in population at the end of the war. By June 1920 the U. S. Housing Office had turned the land and building back to the Navy.

The structure had a number of uses, but in 1936 work began to convert the old hotel into a recreation center. Economic Recovery (Navy) funds paid for the remodeling; a Washington State WPA project, sponsored by the City of Bremerton, provided the labor. The glass roof of the former hotel lunch room was raised and the area converted into a gymnasium/dance hall. A pool and billiards room, rifle range, lounge, reading room, study room and refreshment counter were provided.

A Navy Relief Carnival on May 4 and 5, 1937, officially

This 1936 aerial shows two major Navy Yard additions. On the left is part of the new Machine Shop (Bldg. 431), the largest west of the Mississippi River. At the end of Pier 6, the new Hammerhead Crane partially obscures USS LEXINGTON. Also present at Pier 6 are USS WEST VIRGINIA and USS CALIFORNIA; at Pier 5 are USS MISSISSIPPI, USS MARYLAND, USS PENNSYLVANIA, USS TATNUCK and an Omaha Class light cruiser; at Pier 4 are the KEARSARGE and the USS MOHOPAC. Just leaving the terminal is the ferry CHIPPEWA. Larry J. Jacobsen

opened the Center. On May 12, four days after Captain J. J. London became the Acting Commandant, the building was named Craven Center.[28]

A change occurred in 1938 when the Industrial Dispensary moved into Building 445. This activity had been located in Building 56 at the Bremerton Gate since early in the 1920s.[29] Efforts to obtain the new building reached fruition in 1937 after the Surgeon General visited the Yard and the Chief of the Bureau of Yards and Docks, Rear Admiral Ben Moreel, promised his support.

Captain Gayler returned in 1936 for his third tour as Public Works Officer. The Commandant, Rear Admiral Edward B. Fenner, decided it was time the Yard had a chapel. As he had with the officers' club in the early 1920s,[30] Gayler was able to construct a first class building with a minimum of funds. Bricks and lumber salvaged from other buildings formed the foundation and walls.

Roof beams came from the stable, roof tiles became available when another project couldn't use them. Oak and dogwood trees, shrubs and sod were transplanted from various places around the Yard, and gifts provided the furnishings. The chapel opened in 1938.[31]

For years, the Yard had campaigned for another dry dock. In April 1935, a cruiser-graving dry dock was authorized for the Yard, but funds were not appropriated until June 1936. In the meantime, the Navy Department decided docks capable of handling any size ship were needed on the Pacific Coast. There was much debate as to the type of dock that should be built at Puget Sound. In October 1938, Fenner wrote to the Chief of Naval Operations,

The Navy Yard Hotel was not completed until after the end of World War I. Known later as Craven Center, to honor a former Navy Yard commandant, the rambling building straddled Park Avenue between Burwell and Fourth Street. In addition to the Fourth Street entrance shown here, a tunnel entrance provided direct access to the Yard. This building served a number of purposes before it was demolished in the late 1960s. The Enlisted Men's Club, Park Avenue and a parking lot are now located on the vacated property.

Puget Sound Naval Shipyard

Conversion from power poles to an underground electrical distribution system started with trenches. This 1932 picture shows the Brig (Bldg. 198) and provisions storehouse (Bldg. 197) on the left with the coal bunkers across Farragut Avenue near the Charleston Gate.

Kitsap County Historical Society

Yard personnel in need of treatment reported to Building 56 until the new dispensary (Bldg. 445) was built in 1938. Building 56, formerly the Labor Board, had been moved from its waterfront location to this site near the Bremerton Gate shortly before the construction of Dry Dock 2. Puget Sound Naval Shipyard

expressing his opinion that it should be the larger size dry dock. He pointed out:

> *. . . fifty feet of water can be carried right up to the dock and any vessel built or projected can come here even when seriously damaged and drawing more than its usual depth. There is no other feasible site on this coast which has this immense natural advantage . . . . a big dock can take a small ship, but a small dock cannot take a big ship.*[32]

His arguments were persuasive and the full sized dock was planned. General Construction Company of Seattle received the contract to build the main structural concrete body of the dock at a price of $2,090,900. Later contracts brought the total construction price of the 998-foot long, 132-foot wide, 45 foot deep dry dock and its appurtenances to $4,500,000. The dock was completed late in 1940.

In August 1939 the Yard received authorization to build a fifth dry dock. When completed, its dimensions were similar to those of Dry Dock 4. The placement of Dry Dock 5 required the relocation of Building 50.

Following Hitler's invasion of Austria in May 1938, Congress passed the Navy Expansion Act which provided for a major increase in our country's ships and planes. When Germany invaded Poland in September 1939, President Franklin D. Roosevelt declared a Limited National Emergency.

When Great Britain entered the war, Germany began to mine shipping lanes off the coast of Britain. They used a new type of mine with a magnetic detector which defied the usual sweeping and removal methods. The United States began a crash program to provide its ships with electromagnetic coils to cancel the effect of the steel hulls on the mines' exploders, a technique named degaussing.

The coils were made of multiple strands of several sizes of wire so proper current could be obtained without excessive waste of energy. As the insulation on the early cables was too brittle for handling on a reel, long files of sailors carried the usually 1,000-foot-plus cable slung over their shoulders. The procession wound like a snake from the

Painted white and surrounded by oaks and evergreens, the chapel, built in 1938, on Whiting Avenue, presented this appearance for many years. A social room was added across the front in the mid-1960s. Puget Sound Naval Shipyard

This 1938 photo, looking northeast, was taken from a crane on the east side of Dry Dock 3. The dark building in the foreground was Building 91, Storehouse and Sail and Riggers Loft. Next to it stood the U-shaped Supply Department Building 157, razed to make room for the new Shipfitter Shop, (Bldg. 460), The Fire Station, (Bldg. 346), was located seaward of Building 157. Puget Sound Naval Shipyard

A lowering sky, dappled waters and two small boats add artistic touches to this 1936 photograph of USS SARATOGA. In the left background is Silver Sprint Brewery in Port Orchard. On the right is USS ARIZONA. Larry J. Jacobsen

assembly shop to the ship.

On the completion of work in the Navy Yard, the ships were moved to a deperming station established off shore, north of Illahee State Park. Here the magnetism induced in the ships by direct current welding was reduced. Final testing was done at the Degaussing Station off Jefferson Point, south of Kingston.

William C. Spiller[33] and Rex McIraith[34] were in charge of the work, which required a significant increase in the Electric Shop's work force. An annex was added to the shop for the handling and installation of these massive coils.

At that time approximately 6,000 Puget Sound Navy Yard employees were engaged in the repair and overhaul of battleships WEST VIRGINIA and COLORADO and cruisers MINNEAPOLIS, ASTORIA and HOUSTON. Two yard tugs, ALA and WOBAN, and the 1,620-ton destroyers CHARLES F. HUGHES (DD 428) and MONSSEN (DD 436) neared completion.

Germany invaded France in April 1940.

Additional construction and modernization tasks were assigned PSNY. Keels were laid in October 1939 for two seaplane tenders: BARNEGAT (AVP 10) and BISCAYNE (AVP 11). The following February, CASCO (AVP 12) and MACKINAC (AVP 13) were under construction. During 1940, authorization was received to build five YSDs (self-propelled Seaplane Derricks) and eight 2,100-ton destroyers. The keels for the first four of these destroyers were laid in Dry Dock 3 in March 1941. Then, in December, the Yard received an order to build eight destroyer escorts for the British Navy.

Because of the increase in required design work, a new military billet was established in the Planning Department. Commander Roy T. Cowdrey became the first Design Superintendent.[35] Chief Draftsmen Reuben Johnson[36] and Henry Schairer[37] recognized the need for additional engineers and naval architects for the expanding work load.

Their request to the Civil Service Commission brought 84 junior engineers to the Navy Yard by June 1939, increasing the Design Sections' manpower 50 percent.[38]

Extensive plans were made for modernization of both SARATOGA and LEXINGTON, but the uncertain international situation caused concern at having two of the Navy's eight carriers out of service if the United States became actively engaged in the war then raging in Europe.

To speed work on SARATOGA, an officer and 12 draftsmen from the Navy Yard reported aboard the carrier in San Diego on October 28, 1940. During the trip north, the men began the planning and design work necessary for overhaul and modernization.[39]

As war expanded in Europe, defense measures were initiated throughout the country. In Kitsap County, as elsewhere, certain areas were placed off limits because of "military practice". There was talk of placing a submarine net across Rich Passage and blackouts were planned.

Kitsap County got its first real look at war in the summer of 1941 with the arrival of the British ship HMS WARSPITE. The horrors of war were sealed in spaces damaged by Axis planes; her battered hull was shored against the sea by 12-inch timbers. H. D. Finch[40] wrote:

> *. . . there was a time when we could stand on a dock west of the old administration building and gaze down into the below decks of HMS WARSPITE. In the Mediterranean the ship had experienced a direct hit. The bomb blew out most of her starboard plates to expose much of her inboard. The old ship had traveled halfway round the world to the safe haven of Bremerton. It was also our first view of radar-antennae . . .* [41]

Clyde Caldart,[42] an apprentice shipfitter assigned to the WARSPITE, remembers:

> *. . . Especially wary after so much combat activity, the skipper insisted on isolating the pier where she was moored. A barrier was erected at the inshore end of the pier and no one was permitted entry. That stand-off was shortly resolved by limiting access to those with special badges, and the much-needed on-board renovations proceeded.*

British officers and enlisted men assemble on the bridge of HMS WARSPITE at Puget Sound Navy Yard in 1941. Battle damaged and bedraggled, WARSPITE was a curiosity and heroine to citizens of a country not yet at war. Picture is inscribed by Lieutenant Denys Gregory, RN.

Kitsap County Historical Society

The site chosen for Dry Dock 5 required the relocation of Building 50. Similar in size to Dry Dock 4, Dry Dock 5 looked like this in October 1940. An extension added in 1955 increased Dry Dock 5's length. Puget Sound Naval Shipyard

The ship's officers and men were invited into homes throughout the community, and new friendships were formed. Former WARSPITE sailor Doug Cooper, visiting friends in the area in March 1986, told a BREMERTON SUN reporter:

> *I vividly remember when they pumped out the drydock, the bottom was just covered with salmon. The workers packed them in ice and gave them to our galley crew. That was a real treat. The food on a battleship at sea in a war is, well, not very appealing. Coming to a country that wasn't at war was a completely unique experience for us. I guess people thought we were some kind of heroes because we'd been at war.*[43]

On September 16, 1941, the Puget Sound Navy Yard was 50 years old. The BREMERTON SUN dedicated that day's newspaper to all who had played a part in the Navy Yard's growth. There were articles on the Navy Yard's facilities and history, and congratulatory messages in advertisements from Bremerton businesses. There was no mention of an anniversary celebration being held within the Yard's fence.[44]

The BREMERTON SUN did comment:

> *Due to restrictions imposed by the national emergency, breaking the ground for Drydock No. 5, now under construction in Puget Sound Navy Yard, was not marked with the colorful ceremonies with which construction of Drydock No. 2 was begun here in 1909.*

Articles and editorials in the BREMERTON SUN throughout the fall of 1941 revealed concern that the United States would soon be fully embroiled in the war.

1

Chapter 5

World War II

1941-1945

At 10:25 a.m. Pacific Standard Time, Sunday, December 7, 1941, Japanese planes bombed, torpedoed and strafed the U.S. Pacific Fleet at Pearl Harbor. Repeated waves of Japanese planes attacked U.S. military installations on Oahu, Territory of Hawaii.

Kitsapers heard the news in various ways. Jack Alguard[1] was in the Machine Shop working with Foreman Fred Hicks[2] on an emergency job to straighten a tug's bent propeller shaft. As they finished the work:

. . . someone rushed into the shop with the news that Pearl Harbor had been attacked by Japanese planes. We were aghast, of course, and stunned as we thought what the news portended. I walked out on Pier 6. There a few ships were tied alongside and all was ominously quiet. It was like the calm before a storm. Then, as I returned to the shop, I saw men coming in to work.

It was not yet noon and no call had gone out. None of these workers knew what was needed but they were ready to do something. It was a most strange situation, the air seeming charged with tension at the enormity of the catastrophe and the implications for the long term future. Our world changed dramatically that one spellbound morning, never to be the same again.

Jim Downey[3] heard the news during a sandlot football game at Suquamish:

A U. S. Navy truck pulled up to the field and several Shore Patrol men got out and called our game to a halt. They asked if any of us were in the Army, Navy or Marines, regular or reserve. About four of the players indicated they were Navy Reserve. The SP's said, "Not any more, you are now regular Navy and are ordered to report for duty immediately." Of course we all asked what was wrong and it was then we learned of the Pearl Harbor attack.

On Monday, December 8, the U. S. Congress declared war on Japan.

For workers who lived in Seattle, that Monday brought the first of many changes in accustomed commuting routine, Richard Linkletter,[4] on the ferry KALAKALA's deck, was surprised to hear gunshots and found a rifleman:

. . . firing shots from a standard 30 caliber weapon frequently into the water ahead of the ship to detonate any acoustic mines that might have been laid . . .

Later, we had a submarine net for delay. This open mesh weave of steel cable was stretched on stacks of piling from Bainbridge Island to Orchard Point across Rich Passage and was intended to stop unauthorized entrance of any large craft, surface or submerged. Two barges formed a part of the net and provided a double gateway to all passage of authorized vessels. The ferry

Facing Page

One of the Yard's best kept secrets in 1943 was the pre-fabrication of a new bow for cruiser NEW ORLEANS. When the battle damaged ship arrived in May, both a dry dock and new bow were waiting. Supervising preparations for the match-up is Leadingman Rigger George Wofford. Standing on the new bow and waiting to assist is, third from the left, Shipwright Herbert Abbs. Puget Sound Naval Shipyard

The view from a pier in Clam Bay, Manchester Fuel Depot, shows the open submarine net, an aspect of defense new to the area. Car and passenger ferry WILLAPA has eased through the first opening in the net and waits to negotiate the second exit. The view looks east to the south end of Bainbridge Island. Puget Sound Naval Shipyard

would approach, give the signal, be recognized and the first gate opened by slacking off the line supporting it between the two barges. The ferry entered the space between the barges, and the gate was closed behind it. Then the second gate opened and the ferry proceeded on to Bremerton.[5]

The stunning success of the Japanese attack at Hawaii spread panic on the mainland. The West Coast braced for expected air attacks and invasion. Gas masks, protective clothing and "What to do in case of an air raid" instructions were issued.

The Yard reacted quickly. Reporting to work, Clyde Caldart found WARSPITE turned around to face seaward into Sinclair Inlet, "At four p.m. each day, every shore-connected line was disconnected, and the damage control deck buttoned up for battle conditions."

Barges in front of the dry docks, and nets, which stretched from pier to pier, provided protection from possible torpedo attack. Barrels of water, boxes of sand and rakes were readied at fire watch stations. Criss-crosses of tape appeared on windows, and tall cement-block walls were built for blast shields around buildings' lower floors.

Buildings and storage tanks were camouflaged. Smudge pots and smoke generators on streets and barges were readied to cover the Yard with smoke in response to an air raid warning; blackouts were enforced throughout the area.

Air raid shelters were set up in the basements of Buildings 290 and 467; obsolete boilers from old ships were cut in half and set up to provide some protection for those unable to reach other shelters. In the hill below the officers' quarters, work began on a 1,510-foot tunnel[6] capable of sheltering 7,000 people. The hill near the Marine Barracks provided shelter for military personnel not assigned specific air-raid duties. Contingency plans were drawn up for complete destruction of the Yard to prevent it from falling into enemy hands.[7]

Lieutenant William F. "Pete" Petrovic[8] was Assistant Hull Superintendent for the overhaul then in progress on USS COLORADO (BB 45). Years later, when he returned as a Rear Admiral in command of the Yard, Petrovic told BREMERTON SUN reporter Jerry Grosso;[9]

Administration Building 78 takes on a defensive appearance with criss-crosses of tape placed to lessen injury from flying glass. A cement block wall near the lower floor serves as a blast shield. An air raid shelter stands within the angle of the block wall.

Puget Sound Naval Shipyard

. . . We didn't know the extent of damage done at Pearl Harbor, but as the only battleship repair yard on the West Coast, we knew we would have work to do as soon as the damaged ships could get under way . . . This was the first time that aircraft had been used in anger against battleships. It sure stimulated the installation of additional antiaircraft guns on the COLORADO and other ships.

. . . There were some "scares" during those first weeks after Pearl Harbor . . . A certain number of people would hear a rumor, pack up food and belongings and go over the mountains to Eastern Washington . . .

The anticipated Japanese invasion of Kitsap County did not occur, but the Navy-oriented communities soon felt they had been invaded — by the Army. Soldiers came to defend the Navy Yard and guard transportation routes. In the Yard, the Army manned antiaircraft installations on the roof of Building 290. Guns were installed to cover the Naval and Charleston Avenue gates. Parks, playfields and even backyards were dotted with Army tents.

Bernie Schureman[10] found:

Kitsap County was, essentially, an integrated fort with civilians interspersed among various antiaircraft and balloon sites, motor pool and motor maintenance companies, searchlight batteries and infantry camps.

The Army's barrage balloons were moored on both land and water. Grosso, who grew up in Port Orchard, wrote in his memoirs:

. . . Seems like there always was a new barrage balloon story. Once in awhile one would break loose. "Old Mr. Whomever got out his gun and shot down the one coming toward his house yesterday" was one of the favorite stories, often circulated with just the name of the shooter changed. Then it seemed like the folks on Annapolis Hill had a tremendous lot of trouble with the escaped balloons, because after every windstorm the story went around that one wrapped its long dangling tether cable around some householder's chimney and yanked it right out of the house.

One of the many barrage balloons, a deterrent to dive bombing, floats above a smoke screen created to conceal ships' positions. Vessel in center of photo may have been decommissioned mine layer AROOSTOOK, which the Army used for smoke pots and to anchor a barrage balloon. Puget Sound Naval Shipyard

Six Marine sentries guard the Main Gate while barrage balloons protect the industrial area from possible air attack. The new chain-link fencing provided better access control into the Yard. This May 1942 photo shows Building 290 and, in the background, the Sheet Metal Shop. Puget Sound Naval Shipyard

Part of a Scotch-type boiler, cut in half and turned on its side, becomes an air-raid shelter. This style boiler, previously removed from old ships, filled a need as a temporary shelter. Puget Sound Naval Shipyard

The air-raid tunnel in the hill under the officers' quarters was still incomplete in February 1943; it had to be equipped with benches, sanitary facilities and emergency supplies of food and water. Sick Bay and Tactical Headquarters were to be set up in separate rooms. The shelter measured 1510 feet long, 20 feet wide and 14 feet high in the center. Four separate entries provided access from Farragut Avenue. Puget Sound Naval Shipyard

These Riggers departed for Pearl Harbor the day after Christmas in 1941 under Quarterman Ward Rowell, mainly to remove the big guns and other heavy equipment from the sunken and damaged battleships. They are: back row: Fred Timmerman, Joseph Novak, Peter Wakefield, Ray Server, Palmer Erickson, Carl Miller, Ernest Gibson, Paul Fors and Ward Rowell. Front row: Alexander Brown, Eugene Mort, Hannes Carlson, Harold Rich, Anthony Scrima, Tom Jarvis and Charles Mayer.

Fred Timmerman

The first PSNY employees to see the destruction caused on what President Roosevelt called, "a date that will live in infamy,"[11] were 32 men sent to Pearl Harbor Navy Yard to help in the clean-up and salvage. The group, consisting of 15 machinists under Quarterman Sam Hall and 15 riggers under Quarterman Ward Rowell, was on its way by December 27.

Rigger Fred Timmerman's[12] memoirs of the occasion explain:

> *We reported to our respective shops and were quickly assigned to work with our counterparts. The riggers used a 100-ton floating crane . . . to lift armor from the turrets and remove interference and equipment necessary to reach the guns. Our first ship was WEST VIRGINIA. I also worked on ARIZONA, removing guns from her No. 3 turret.*

As much weight as possible was removed from the ships in order to ease the task of raising the vessels. The guns were used elsewhere. Two more groups of PSNY employees left for Pearl Harbor during January. These included gas cutters, burners, shipfitters, caulkers, drillers, punchers, shearers, welders and loftsmen.

The first of the Pearl Harbor Ghosts[13] arrived on December 29, 1941. This was USS TENNESSEE (BB 43), her after-section burned and blasted by bomb hits and scarred by fiery debris from USS ARIZONA. TENNESSEE was followed the next day by USS MARYLAND (BB 46). These two were the least damaged of the battleships attacked at Pearl Harbor. The people of the Puget Sound Navy Yard, already busy with new construction, began to realize the extent of the job which lay ahead.

The fleet desperately needed TENNESSEE and MARYLAND. New construction was put on hold while workers toiled day and night to send the ships to sea. In addition to repairs to her after section, TENNESSEE received new

A bomb penetrating NEVADA's upper decks, exploded when hitting the armored deck below, causing the pictured destruction. A hatch and the fo'c's'le deck were blown upward and out of place alongside the main battery turret. The teak planks of the deck were distorted but, for the most part, not badly splintered. At PSNY, NEVADA received entirely new teak decks. Donald K. Ross

USS NEVADA, after repairs and conversion, is rigged for deperming at a barge off Illahee. Energized cables, wrapped vertically around the ship's hull, provided power for the deperming to reduce the ship's residual magnetism. After deperming, ships proceeded to the degaussing range at Jefferson Head near Kingston for calibration of their degaussing systems. Puget Sound Naval Shipyard

USS TENNESSEE stands out for sea trials in May 1943 after her second battle damage repair availability and modernization. New radar antennas are seen at the top of the foremast and on the Mark 37 gun directors, port and starboard. New five-inch 38 cal. twins, 40 mm and 20 mm antiaircraft gun batteries are visible amidships on the starboard side. The boat and airplane crane still stands on the stern. Puget Sound Naval Shipyard

14-inch rifles[14], her antiaircraft guns were removed, and 20 and 40 mm batteries were installed. Below decks, 130 watertight compartments were constructed.

MARYLAND's after section was completely rebuilt; additional steel for splinter protection, blisters and 745 tons of armor plate were added. She also received eighteen quad 40 mm gun mounts for antiaircraft protection. The two ships left to rejoin the fleet 53 days after their arrival.

In February 1942, SARATOGA returned to PSNY, only two months after an extensive overhaul here,[15] SARA had taken a torpedo hit on the port side amidships on January 11 and suffered extensive hull and machinery damage. Her repair and modernization were given top priority during the ship's three-month availability at PSNY; the need for carriers at the battle front was crucial.

In addition to the repair of battle damage, SARA's old antiaircraft battery and fire control system were removed and replaced with new 5″/38, 40 mm and 20 mm gun systems. A fuel oil filling and stripping system was installed in a new starboard blister.[16]

The third of the five "ghosts" repaired at PSNY, USS NEVADA (BB 36), arrived May 1, 1942. She had been the only battleship to get under way during the attack, even though she was down by the bow from a torpedo hit. Enemy planes pounded NEVADA, forcing her to run aground on the Waipio Peninsula. As would be the case with most ships during the war, ship's officers and men worked alongside Yard workmen to prepare the ship for its trip to Puget Sound.

The Yard spent more than 700,000 man hours during the next seven months preparing NEVADA to fight again. The time was well invested, for NEVADA waged war from Massacre Bay in Alaska to both shores of France and back to the Pacific.

TENNESSEE returned to PSNY in August 1942 for a nine-month availability. In spite of repairs and improvements made earlier, TENNESSEE was still an old battleship. Urgently needed during the first months of the war, TENNESSEE fought well; now she could be spared to be modernized.

Before TENNESSEE'S second departure in May 1943, USS CALIFORNIA (BB 44) and USS WEST VIRGINIA (BB 48) arrived. Although both ships made the trip under their own power, their condition was a startling contrast to TENNESSEE. Sunk in Pearl Harbor's silt and sticky mud, and completely flooded, WEST VIRGINIA and CALIFORNIA were the most heavily damaged of the "ghosts" that came to PSNY. Both ships were stripped to their second decks and rebuilt. These five battleships created a heavy workload. An average of 2,971 people per day worked 30 months to return them to action.

Before any of the ships arrived, the Pattern Shop had searched out patterns and molds developed during earlier overhauls, thus greatly expediting preparation for the repair work. Information compiled earlier by the Electric Shop aided electricians in rewinding the large water-damaged electric motors. Fred Johnson[17] of the electrical test section explained:

SARATOGA shows her new camouflage war paint as she leaves the Yard in September 1944. SARA received repairs of storm damage she received near Tasmania in March. Flooding in her elevator pit caused electrical damage and fires.

Puget Sound Naval Shipyard

This January 1944 photograph of USS CALIFORNIA shows the antennas for her new radar systems: antiaircraft fire control radar at the left of the picture, large air search radar and surface search radar at the top of the foremast and main battery fire control radar on the right.

Puget Sound Naval Shipyard

Shown here in Dry Dock 5 in May 1943, part of NEW ORLEANS' temporary bow has been cut away. The rest of the jury-rigged bow will be removed and replaced with a welded pre-fabricated bow prepared by PSNY before her arrival. Guns have been removed from Turret No. 2 and gun ports are blanked. Puget Sound Naval Shipyard

Every piece of equipment that was tested had its nameplate data recorded and had the information what was done to it. If the coils or armature were rewound, the information as what size wire was used and number of turns put on [was noted] . . . *This information came in very handy when our fleet was sent to the bottom in World War II.*

In the darkness of midnight November 30, 1942 at the Battle of Tassafaronga, a torpedo ripped into the port bow of USS NEW ORLEANS (CA 32), "abreast two magazines. These united with the torpedo blast to rip off the forward part of the ship as far back as No. 2 turret."[18] Minus her entire bow and with her engines in reverse to lessen pressure on shored-up bulkheads, the ship made her way to Sydney, Australia. With a temporary bow installed there, NEW ORLEANS eased her way to Bremerton.

Under strict secrecy,[19] a new bow had been built for NEW ORLEANS, using plans from her sister ship ASTORIA. According to PACIFIC NORTHWEST GOES TO WAR:[20]

. . . when the 8,000 tons of ship was nosed into the 1,600 tons of new bow, it was found that it was out of line only one-eighth of an inch.

The story is often told that when Hull Superintendent Commander Emmett E. Sprung[21] checked the bow, he pretended great disappointment, but then laughed and rejoiced with the others at the success of the fit. NEW ORLEANS was soon as good as new, but differed from other ships in her class because her bow was welded[22] whereas the rest of the ship had been riveted.

USS ABNER READ (DD 526) lost her stern in an underwater explosion off Kiska, Alaska, in August 1943. Puget Sound Navy Yard had a new stern waiting when she arrived in October under tow of tug ORIOLE.

USS ENTERPRISE (CV 6) arrived at the Yard July 22, 1943, for overhaul and battle damage repairs. Approximately 369,300 man hours were spent in the next three months to return her to the fleet. Petrovic was the Ship Superintendent over-seeing the installation of blisters and gasoline saddle tanks on this battle scarred carrier.

Aircraft Carrier BUNKER HILL was the victim of two bomb-carrying Japanese suicide planes, which struck the flight deck. In the upper photo, workmen survey damage caused by the second plane. Below, the damaged structures have been cut away and repairs are under way.

Puget Sound Naval Shipyard

Not only enemy action, but weather brought grief to ships. While enroute to Tasmania in March 1944, SARATOGA was in a severe storm and suffered damage to her electrical system when the elevator pit flooded, necessitating a trip to the Yard for repairs. Buck Wynn,[23] who worked on the ship, remembers that almost the entire hangar deck required sandblasting and painting, because of black soot from the electrical fires.

Another weather victim was USS PITTSBURGH (CA 72). Caught in a typhoon with winds up to 150 miles an hour in June 1945, she lost over 100 feet of her bow. Fitted with a temporary bow in Guam, she made her way to Bremerton under her own power. As with other ships, Master Shipfitter Herman Petersen[24] and his crew had a prefabricated bow waiting.

During the less than four years of World War II, the Puget Sound Navy Yard repaired sixteen battleships, nine heavy carriers, four light carriers, four heavy cruisers, five light cruisers, 67 destroyers, four submarines and 78 auxiliary and small ships.[25] Some of these ships were at PSNY two or three times for repairs during the war years.

SARATOGA required repairs at PSNY three times during the war. She was at PSNY again from March 17-May 14, 1945, for repair of massive damage suffered when six kamikaze planes hit her.[26] The Chief of the Bureau of Ships, on June 19, wrote to the Commandant, officers, men and women of Puget Sound Navy Yard:

> *The record breaking repair job you performed on USS SARATOGA reflects the highest credit on the Puget Sound Navy Yard. Although the SARATOGA was more extensively damaged than any vessel previously received at the Yard, you were successful in accomplishing one of the fastest repair schedules ever attempted on a capital ship. By working night and day, effecting ingenious shortcuts and utilizing all the skill and resolution traditional with PSNY, you have rebuilt this veteran carrier with remarkable speed and sent her back into action in fit shape once more . . .*[27]

USS PITTSBURGH enters Dry Dock 2 in July 1945 after losing her bow in a Pacific Ocean typhoon. Her new bow, prefabricated using hull plans and templates available in the Shipfitters Shop, was waiting for PITTSBURGH's arrival. Workmen immediately removed the temporary bow and installed the new one. The war ended before she was ready for sea.

Puget Sound Naval Shipyard

This view taken from the hammerhead crane is typical of the wartime pier congestion. North of Piers 6 and 7 are the Power Plant, Dry Dock 3, and the Shipfitters Shop. Puget Sound Naval Shipyard

The blizzard of January 1943 did not stop production at the Navy Yard. Public Works worked around the clock clearing roads, disposing of snow wherever possible and washing streets with salt water. Large crews of Navy men cleared the highways while Army trucks pulled buses out of heavy drifts. Puget Sound Naval Shipyard

The first group of four Lend Lease Destroyer Escorts was laid down on the building ways southwest of Dry Dock 5 in September 1942. The keels, resting on the blocks in this picture, were for BDE's 37, 38, 39 and 40.

Puget Sound Naval Shipyard

TABLE OF EVENTS

★

Mrs. Cecily V. Reddin, Sponsor

★

Oper. No.	Pacific War Time
	0800—Hoist colors; full dress ship.
1	1400—Launching crew reports. 1430—Launching rehearsal.
2	1517—Test Jacks.
3	1642—Remove distance pieces. Remove #2 cribs.
4	1702—First rally — red wedges.
5	1722—Second rally — white wedges.
6	1742—Third rally — all wedges.
7	1747—Remove ram rails, spur shores. Remove dagger shores. Remove red blocks. Personnel assigned to ship take station.
8	1827—Remove white blocks and #1 and #3 cribs.
	1842—Remove forward dagger shores. Inspection between ways. Flag hoist on crane to clear channel.
	1845—Selections by Yard Band.
9	1850—Report on final inspection.
10	1855—All clear of building slip. Brief remarks by Commandant.
	1905—Presentation of bond to Sponsor by Manager.
	1910—Prayer, followed by National Anthem.
11	1915—Remove trigger shores. Remove trigger locking pin. 1 long blast on horns. Trip dog shores. Ten seconds standby signal.
12	1917—Launch.

P.S.N.Y. 5-14-43 1M

During World War II's frantic days, simple Tables of Events for a ship's launching substituted for the usual elegant programs.

Puget Sound Naval Shipyard

The Yard repaired, modernized, converted, built and fitted-out ships. "Fitting-out" is the installation of operating equipment in newly constructed ships. During 1942 and 1943 the Yard fitted-out 191 vessels. Less fitting-out was done later when PSNY helped private contractors to establish their own facilities. This allowed the Navy Yard to use its limited manpower for repair, conversion and new construction.

The Escort Carrier (CVE) Conversion program had been started under the Lend Lease Act to combat the alarming increase in ship sinkings by German submarines during the Spring of 1941. With our entry into the war, even more of these escort vessels were needed.

The Yard converted into CVEs five Maritime Commission type C3 cargo vessels, on which construction had already begun at Seattle-Tacoma Shipbuilding Corporation's Tacoma Yard. The first, USS COPAHEE (CVE 16) arrived February 8, 1942, followed by USS NASSAU (CVE 17) and USS ALTAMAHA (CVE 18). These conversions were completed in June, August and September that year. Later, the Yard converted PRINCE WILLIAM (CVE 31) and GLACIER (CVE 33). The main mission of these "jeep" carriers was to provide air cover for protection of convoys.

The Yard also converted five over-age destroyers, WATERS, DENT, SCHLEY, WARD and RATHBURNE, to APDs, high speed troop transports. LONG ISLAND (CVE 1) became an Airplane Transport vessel.

Throughout the war, new construction was delayed by the priority of repair work and the difficulty of acquiring needed materials. Keels were laid in Dry Dock 3 on June 3, 1941, for HALFORD (DD 480) and LEUTZE (DD 481), the first of eight destroyers authorized in 1940. HALFORD was completed in May 1943. LEUTZE, HOWARD (DD 592), KILLEN (DD 593), HART (DD 594), and METCALF (DD 595) joined the fleet during 1944. Because the heavy load of repair work required almost all of the work force in the fall of 1944, SHIELDS (DD 596) and WILEY (DD 597) were not completed until the spring of 1945.

USS GREINER (DE 37), the first of the Destroyer Escorts launched from the new destroyer building ways, floats free of her launching cradle on May 20, 1943. Shops with the lowest number of absentees during a specific week had the honor of choosing sponsor and flower girl for christening ceremonies for the destroyer escorts. In the winning Outside Machine Shop, E. O. Reddin's badge number was drawn. He selected Mrs. Cecily V. Reddin, a cafeteria employee as sponsor; Lowell Anderson of the Foundry selected daughter Gloria Jean as flower girl for Greiner. *Puget Sound Naval Shipyard*

Officers and crews assemble east of Building 433 for ceremonies observing completion of CVE Pre-Commissioning School in August 1943. Under Commanding Officer Lieutenant Commander John W. Collier, these men trained for duty on the small carriers. In the background are Marine Officers' Quarters and Navy Chapel. Kitsap County Historical Society

A major factor contributing to the delay in constructing the destroyers was the higher priority given to destroyer escorts. Shortly after work began on destroyers, the Navy Yard was assigned construction of eight destroyer escort vessels, BDEs 37-44[28].

These were eight of the fifty ships ordered built for the British Navy under the Lend Lease Act. Keels for the first four were laid on September 7, 1942, on newly constructed building ways.[29] The following January keels were laid for the other four ships in Dry Dock 3. All eight were completed by the end of 1943. Because we were now at war, the ships were not transferred to Britain, but were redesignated as DEs and joined the U.S. fleet.

The names assigned to the ships were GREINER (DE 37), WYMAN (DE 38), LOVERING (DE 39), SANDERS (DE 40), BRACKETT (DE 41), REYNOLDS (DE 42), MITCHELL (DE 43) and DONALDSON (DE 44). Completion of these ships required considerable pre-fabrication and farming out.

Although the Yard's "Farming Out" program had been used before, its use became imperative in March 1942 due to the urgency of the destroyer escort program. Lieutenant Commander Charles E. Trescott, Farming Out Officer, was assisted by other officers, Civilian Administrator Russell Sweaney, four estimators and up to 20 clerks.

Subcontracts were let throughout the Northwest for

construction of needed material, e.g., piping, electrical appliances, doors, valves, structural shapes. The government supplied participating firms with necessary raw materials, tools and plans. The Yard assembled and installed the prefabricated items.

Norma McCleary (Pappas)[30] set up the filing system to keep track of the distribution and return of government material and plans. She remembers the busy office, with typewriters clattering and phones ringing:

> *We were a close knit group, sharing each other's joys and sorrows. We'd go skiing together when anyone had enough gas coupons for us to get to the slopes. Lieutenant (jg)s Milan Pittman and Colin McKay often skied with us. They were very capable and well liked and we weren't surprised that they both reached high rank in the Navy.*

The Yard also built nine yard oil and gasoline carriers (YOGs 72-80) and six barracks barges (APLs 2-7), making extensive use of prefabrications from the Shipfitter Shop. The APLs were later completed at a private shipyard to save manpower in the Yard for repair jobs.

Manufacturing tasks were also assigned to the Yard. During 1942-1945, approximately 245,000 valves were made. Most ranged from 1/8 inch to 10 inches in diameter, but some were as large as 24 inches for ships' condensers or five-foot valves for dry docks.

Boat building increased dramatically; during 1943 alone, the Boat Shop constructed 434 boats. One particularly interesting assignment was the construction of transparent plastic ship models.[31] The Yard's pre-commissioning Training Unit requested the Boat Shop to build a ¼-inch scale model of the CVE 54-104 (CASABLANCA) class. The model could be taken apart by decks and bulkheads to familiarize the crews with all spaces during firefighting and damage control instruction. The usefulness of this model led to orders for more models to be used at other training stations.

The Yard's construction and repair records were no secrets within the Yard. Above the Joiner Shop entrance, a large sign enumerated numbers and types of ships built and repaired during World War II. The work force of 32,500 had built 50 ships for an expenditure of $137,500,000 and 5,250,000 man-hours. Repairs had been made to 363 ships.

Kitsap County Historical Society

Long lines of Yard applicants stretched from the Labor Board's entry into Building 290, down the alley parallel to Pacific Avenue in this July 10, 1943, SALUTE photo. An extensive recruiting campaign, carried on nationwide for four months, paid off with a daily deluge of people eager to contribute to the war effort. Puget Sound Naval Shipyard

A tremendous number of people, civilian and military, were needed to do the Navy Yard's job. The Navy Yard set a goal of 36,000 workers but for many reasons was never quite able to reach it.

The accustomed recruitment by local Civil Service and U.S. Employment Service, which hired for skill and experience, failed to fulfill the Yard's needs. Instead of coming to the Navy Yard, local workers with experience sought higher paying jobs in the burgeoning aircraft industry and well established shipbuilding plants in Seattle. A dearth of housing in the Bremerton area, or a long ferry ride from residence, discouraged workers from applying at PSNY. Something had to be done.

At the suggestion of a visiting Selective Service official, two naval officers recruited throughout the state, using movies, radio and press to attract workers. The officers' familiarity with the Yard plus the psychological effect of the uniforms' gold braid added to many successful presentations. Favorable results were soon seen as new workers arrived. However, long and often heated discussions with the Manpower Commission continued, regarding the priority of the Yard in relation to other war industries, to agriculture and to logging.

Efforts to recruit in areas of the country not experiencing labor shortages were unsuccessful because workers were unable or unwilling to make the long trip to Washington State unless they were assured of jobs. The President's Emergency Fund offered the means to take advantage of this situation by providing funds to pay the transportation of recruits to the Yard.[32]

A national campaign to attract 7,500 journeymen and helpers was begun through the Civil Service Commission. A month spent in Minnesota by recruiters resulted in 3,838 workers reporting to the Navy Yard. During the first part of July 1943, as many as 300 persons a day arrived.

Occasionally recruiting campaigns were halted because of insufficient housing. Housing and recruiting agencies were at odds. The U.S. Employment Service refused to recruit from other areas because Bremerton was declared a critical housing area. The National Housing Authority could not authorize home building unless the War Manpower Commission could assure jobs for future residents.

The housing shortage developed before the start of the war. In 1940, the Bremerton Housing Authority started work on West Park Housing,[33] which opened April 14, 1941. "Duration Dormitories" for single men and women

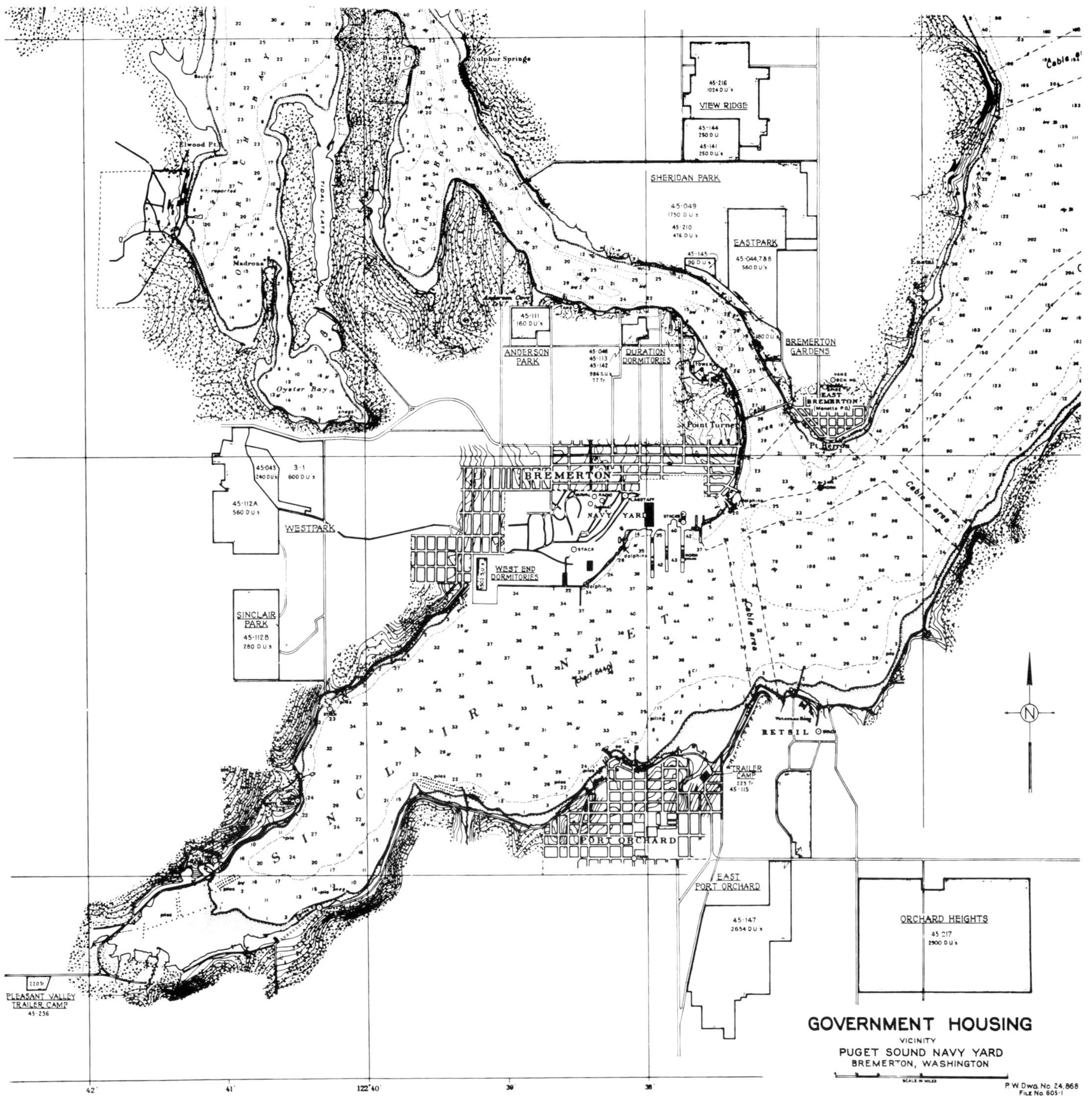

Government housing for Yard employees and Navy personnel was located throughout the local area with units as far south and east as Orchard Heights beyond Port Orchard and as far north and east as View Ridge in what is now East Bremerton.

Puget Sound Naval Shipyard

were built where Olympic College now stands. Later, four more were built in the Yard's west end. The Navy Yard assisted contractors in getting scarce building materials to construct housing units throughout the Navy Yard area.

Trailer camps had sprung up before the War. By August 1942 it was estimated that 2,500 families lived in trailers, often without proper sanitation. All efforts were made to find more housing, individuals were urged to make room for Yard workers in their homes. People no longer working in the Yard had to vacate government housing. When other housing was not available, garages and chicken coops were remodeled. Defense workers slept in cars, tents, basements and attics. Rooming houses featured "hot beds" where three shifts of workers kept the sheets warm.

Ship personnel and their families doubled up, commuted from Seattle or took housing wherever they could find it. Two young officers on USS DEWEY shared temporary housing, moving with their wives and infants into a one-bedroom shack at the end of a pier below the Enetai Inn. When the tide was out they lived above a beach, when it was in, over water. Some military officers and top echelon Yard employees were fortunate enough to be able to rent in privately owned Bremerton Gardens in Manette.[34]

The Navy Yard promoted the construction of government housing. By war's end, there were in Port Orchard, 2,654 units at East Port Orchard and 2,902 at Orchard Heights. On the east side of the Washington Narrows were 1,524 units at View Ridge, 560 at East Park, and 2,226 at Sheridan Park. On the west side were 600 at West Park, 560 in West Park Addition, 160 at Anderson Cove and 280 at Sinclair Heights.

Many black families lived in surrounding towns, but recruits were assigned to Sinclair Heights. Already established blacks assisted the new arrivals. Gertrude Joseph, who lived there, remembered the Reverend Lewis Cooper who was particularly helpful and who persuaded the Navy Yard to provide a building for a church at the project.[35]

Quincy Jones Sr., a carpenter from Chicago, and his two sons arrived at Sinclair Heights on July 4, 1943. Their house was small and poorly heated, but for the boys it was a wonderful place; it was in a wooded area with much to explore and wild berries to eat. The older boy had been interested in music while in Chicago; now he received encouragement from musicians in the Navy Yard and from teachers in Bremerton Schools. The family moved to Seattle in 1947 and Quincy Jones Jr. began his career as musician and producer.

In 1990, Quincy's brother, Lloyd, told the authors:

Moving out here was probably the best thing our father could have done for us . . . I don't know how many black families remained in the Bremerton area, but many of those who lived in Sinclair Heights are here in Seattle now.[36]

Lieutenant Commander O. D. Adams and Lieutenant (jg) Ted Barnowe, both professional educators, were in charge of training; converting farmers, housewives, service station attendants into shipyard workers. As Apprentice School classes were already as large as could be handled, training facilities were set up in all major shops to train helpers and mechanics.[37]

Instructors were enthusiastic about the booklet, "WORKING WITH THE NAVY", which became available the summer of 1943. The compact 76-page book,

Some of East Park, a Navy housing area of duplexes and fourplexes in East Bremerton, is still in use by Navy personnel and their families. Some of these buildings have since been sold to individuals and businesses for private homes or apartments.

Puget Sound Naval Shipyard

A variety of dwellings is seen in this World War II photo taken on Kitsap Way near Wilbert Street. Cars, tents, travel trailers, one-room shacks and genuine houses shelter newly arrived workers. A car wash and a restaurant now occupy this site. Larry J. Jacobson

printed in the Yard by Quarterman Emmet Kidrick's Shop 93, contained a wealth of information for the benefit of the Yard's workers. It had several reprintings and the August 11, 1944, SALUTE explained that the material had been typed on an electric typewriter on a special "Duplimat" plate, and after being run through the presses, the reams of paper were gathered by assemblers, who picked up each page as they walked around a long table.[38]

Praise came from many sources, and thousands of the books were sent to other Yards. Training manuals for rigging, coppersmithing, sheet metal work, diving, marine electricity and shipfitting were available by the summer of 1944, and more were in progress.

The Navy Yard's civilian divers instructed officers and enlisted personnel in the art of diving. In the fall of 1943, six divers and a salvage officer of the Army Engineers Corps graduated from the class. Hundreds of enlisted men qualified as second class divers and received training in underwater welding, gas cutting and burning. Leadingman Rigger W. B. Evertson was in charge and Thomas J. Craig Jr, Theodore R. Donahue and Earl Lawrence were instructors.[39]

The Board of Labor had a difficult task. Often men and families arrived tired, hungry and without funds. The USO Travelers' Aid helped many who arrived without funds to carry them the two weeks until they received their first paychecks. Later the Civilian Aid office was set up, with a loan of $5,000 from the Cooperative Association.[40] This was later supplemented by an appropriation of government funds of about $150,000 and contributions from individuals.

No cash was given out, but workers received chits for credit for food and lodging until payday. The loan was repaid through payroll deductions. Workers of the Civilian

Willing to serve his country in time of war, this partially blind elderly man had driven his goat cart from Idaho to the Yard. Employees Hope Beck and Maxine McDaniels are ready for a ride, while Bob Crial, placement examiner, nurse Mary Osterhaut and another employee look on. Hope Beck

Aid Office interviewed arrivals to discover their needs, counseled them, arranged for credit for services in town, and provided information about the area. Many needed medical attention.

Labor Board receptionist Hope Beck[41] won the hearts of the newcomers with her smile and caring ways. She recalls:

> *. . . if a man was rejected for any reason, he was given a return ticket to his point of origin. But it got so bad, we were so short of workers, he had to be one day from the grave to be rejected.*

When recruiters did not screen the applicants carefully, some would-be employees were found to be completely incapable of holding any job in the Yard. Some merely had difficulties adjusting. Hope Beck remembers:

> *Many midwesterners had never seen a big body of water and had no idea what a dry dock was. One man, sent to Dry Dock 1, wandered back in three hours. "Lady", he said, "I tromped up and down in this Navy Yard and there ain't no such thing as a dry dock; everyone of them has got water in them."*
>
> *Electric ranges were new to some families; one husband built a fire in the oven in his housing unit . . . one fellow tipped me for reading his letter from home to him. I had typed out a reply and put a stamp on the envelope. When I handed back his change from a nickel, he insisted; "Oh no, lady, that's your tip!"*

To meet these hitherto unknown personnel concerns, Lieutenant Commander D.J. "Mike" Sass was appointed Personnel Officer. Bob Doubleday[42] described Sass:

. . . an ex-submariner with an outgoing personality, more than adequate energy and enthusiasm and a liking for people. He entered his job with characteristic verve and took on a number of young officers who were beginning to arrive for duty in the Yard to handle the boatloads of problems that were showing up . . . His civilian assistant was Bernard Burke.[43]

Prior to the outbreak of war, Executive Orders 9250 and 9328 set Navy Yard wages at levels prevailing in private yards in the same area. On April 23, 1941, the American Federation of Labor and the Army, Navy and Maritime Commission signed a "Master Agreement" for the West Coast that established work, overtime and pay schedules. The Bremerton Metal Trades Council, an affiliate of the AFL, represented most of the artisan trades and other non-classified workers at the Puget Sound Navy Yard.[44]

With representatives from all Navy Yards, 20 delegates from PSNY attended a conference called by Secretary of the Navy Frank Knox in Washington, D.C., in October 1942. The purpose of the meeting was to determine ways to attain the greatest possible cooperation between management and labor in order to win the war.

Representing PSNY were Metal Trades Council members C. P. LaGuardia, J. W. Jordon, O. A. Burrows, Sherman Davis, O. G. Brown, C. R. Thomas, Kenneth Payne, William Hansen, Vern Siegner, E. V. Lyttaker, Harry Richardson and Stanley W. Oliver. Representing non-union employees and non-AFL unions were Ray Gundlach, Leonard Wager, Glenn Halladay, Anton Hanson, Eldon Miller, Robert Ross, Alfred Lockness and Muriel Job. George Penketh was the Master Mechanic delegate and Vernon B. Mills represented the Supervisors, while Sass and Lieutenant Commander J. S. Grant represented the military.

The delegates returned full of enthusiasm, Oliver, Secretary/Treasurer of the Bremerton Metal Trades Council, said:

I am sure that all who attended the conference were sharply impressed with the necessity for even greater production. I was especially impressed with the story of Lieut. Commander John D. Bulkeley.

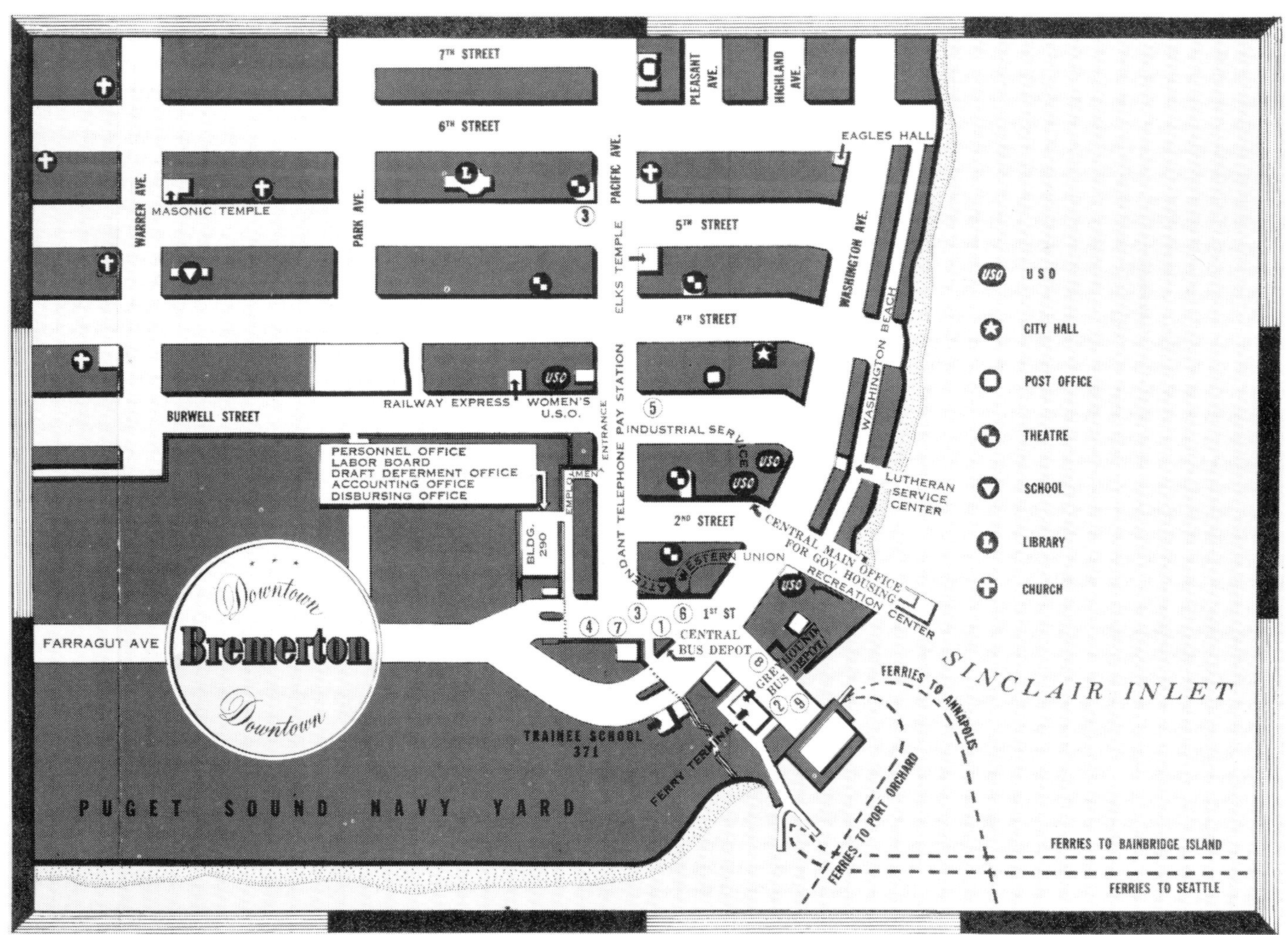

Downtown Bremerton had four theaters, a library and several post offices. Two USOs, a Lutheran Service Center and the YMCA provided social atmospheres for Navy personnel. The Masonic Temple, Elks Temple and Eagles Hall were busy spots for special parties for both civilians and military.

Puget Sound Naval Shipyard

He was also impressed with Lieutenant Commander Frank Szehner, the Assistant Repair Superintendent (Electrical) at PSNY. Oliver's memoirs state:

> *About the time of Szehner's reporting* [to the Navy Yard], *a serious imbalance had developed between the distributions of manpower in various shops and crafts. Recruiting had been to standard ratios for shipfitters, machinists, pipefitters, without allowance for the greatly increased need for electricians on the modern ships to install, repair and replace electrical systems. He soon recognized this imbalance and reported this to top management, resulting in a major change in recruiting and training efforts.* [45]

Many PSNY workers continued their contribution to the work effort after hours. The Coast Guard quickly recognized the value of local yachtsmen's knowledge of Puget Sound waters and organized Coast Guard Flotilla No. 32 with Marx Libby as Commander. Boat crews had regular weekly patrolling assignments on waters inside the submarine net, but were on call at all times for extra assignments.

Serving in the squadron were: A. Mosher, S. E. Hess, Ed Werner, Ken Chase, Oliver Renn, Fred Rabb, Don Edson, Bill Snyder, Paul Engle, George Giblett, L. E. Morris, Clyde Hayes, C. J. Downie, C. G. Titus, Roy Lycksell, L. Chandler, C. H. Dougherty, H. D. Thompson, M. H. Appel, F. M. Svacina, Ted Engbretson, Kermat Olson, Ted Wright, George Sowers, Hank W. Kuhlman, H. B. Garrett, A. C. Anderson, L. A. Tucker, Ed Oates, J. J. Sands, Ed Stell, T. Bontroger, G. Ristow, P. P. Thompson and J. McGinnis.[46]

At a giant bond rally in the Machine Shop, on February 22, 1943, Navy Yard Commandant Sherwood A. Taffinder announced that Secretary of the Navy Frank Knox had approved a six month bond selling competition between the Puget Sound and Mare Island Navy Yards to determine whether the name USS BREMERTON or USS VALLEJO would be given to a cruiser being built on the East Coast.

Mare Island was leading as the drive neared its end, but at a massive rally on Farragut Avenue, film star Ginger Rogers urged workers to buy more bonds. They responded, putting Puget Sound ahead. After a campaign characterized by SALUTE as having "as many ups and downs as a supply building elevator," Mare Island conceded the victory to PSNY. Both Yards had topped their Navy Department-set quota, Mare Island with 123.7 per cent and Puget Sound with 125.2 per cent.

Secretary of the Navy Frank Knox, standing under the "Buy Bonds" canopy is greeted by 10,000 Yard employees at a Bond Rally in July 1943. Knox described PSNY as "the Yard nearest the battlefront." The Secretary set an example for workers by purchasing a bond in the Shipfitters Shop.

Puget Sound Naval Shipyard

Each of the ten Navy Yard groups[48] having the highest percent of gross pay pledged for bonds during the drive submitted the name of a woman employee to sponsor the new ship. At the August 7 rally, Bremerton Mayor L. "Hum" Kean drew the name of the finalist. It was Betty McGowan,[49] representing the Rigger and Shipwright Shop.

Captain of the Yard, Captain Lucian F. Kimball presented bouquets from the Yard Cooperative Association to all the candidates and to Mrs. Juanita Keyes, the Matron of Honor and Cynthia Coyle, the Flower Girl. They were the candidates of the Round Table of West Bremerton, which won a bond-selling contest among the Bremerton service clubs.

Betty McGowan "well and truly" christens USS BREMERTON at Camden, New Jersey, on July 2, 1944. Betty McGowan

USS BREMERTON sponsor Betty McGowan poses with Matron of Honor Juanita Keyes and Flower Girl Cynthia Coyle. With them is their host, Fred A. Cornell of the New York Shipbuilding Corporation. Betty McGowan

The Bremerton Chamber of Commerce spearheaded a drive to raise money for the BREMERTON'S ship's bell as a gift from the city. On January 21, 1944, the foundry began pouring bronze for the 1,250 pound bell. It was heat-treated in the Forge Shop, surface-finished and engraved in the Machine Shop and received final plating and polishing in Shop 51.

SALUTE reported on February 4, 1944, that the tone of the bell would be B-flat, explaining that the size and tone of the bell reflected the size of the ship, and that was the tone for ships of more than 12,000 tons displacement.

On a very hot July 2, 1944, sponsor Betty McGowan and her attendants stood on a platform at the New York Shipbuilding Company Yard in Camden, New Jersey, to break the traditional bottle of champagne across the ship's bow. She was successful on the first swing. USS BREMERTON was christened and launched.

In Bremerton, the day was marked with a flag raising ceremony at Roosevelt Field and a double-header baseball game at which an additional $11,250 in war bonds was sold.

On the Pacific Coast, approximately 150,000 civilians were volunteers of the Aircraft Warning Corps. At the South Colby post, 22 of the volunteers were PSNY workers.

Military and civilians also showed their patriotism in the purchase of War Bonds. Between November 16, 1941 and December 31, 1945, PSNY employees pledged $44,438,175.[47]

The serious problem of transportation during the prewar employment build-up was intensified with the outbreak of war. Tires and gasoline were in short supply and were rationed.

In 1941, 51 motor coaches traveled on 37 daily routes, transporting an estimated 4,500 workers from as far away as Lake Bay in Mason County. Eventually 90 buses, carrying approximately 600,000 passengers, traveled 200,000 miles a month.

Within the Yard, small fleets of bicycles were assigned shops for use by waterfront supervisors. Some shops also had a pickup truck. Two buses shuttled people from one end of Farragut Avenue to the other on a ten minute schedule. During the rush periods at the end and beginning of each shift, buses which had just brought a load of workers from outside were pressed into service.

Supplementing the buses was a three-coach train, the Cannon Ball Express, which made the trip between the Main Gate and the west end three times on the day and swing shifts and twice for the graveyard shift workers. Benches

Buses line up on Farragut Avenue to transport workers between work and home. The structure on top of Building 50 was used as a watch station for air-raids and fire watches. Puget Sound Naval Shipyard

FERRIES

Effective June 2

Seattle-Bremerton

Ferries Kalakala, Chippewa, Willapa and Enetai

DINING SERVICE

55-MINUTE CROSSING

Lv. Seattle	Lv. Bremerton
★6:15 am	6:20 am
6:30	•7:25
7:30	★8:00
8:00	8:40
8:35	9:15
9:10	9:50
9:50	10:25
10:25	11:00
11:00	11:35
11:35	12:10 pm
12:10 pm	12:45
12:45	1:20
1:20	1:55
1:50	2:30
2:30	2:55
3:05	4:00
3:40	4:40
4:00	5:00
5:25	5:10
5:55	6:35
6:35	7:10
7:10	7:45
7:45	8:20
8:20	8:55
8:55	9:30
9:30	10:05
10:15	10:40
11:05	11:25
● 11:00	● 12:10 pm
†12:30 am	12:55

★ Except Sunday
• 7:40 on Sunday
● Saturday Night only
† 1:00 am on Sunday Morning.

Car and driver $1.10 each way
Passengers 45c each way

BLACK BALL LINE

FERRIES

EFFECTIVE MARCH 1

Lv. Bremerton	Lv. Seattle
5:50 a.m.	5:45 a.m.
7:00 a.m.	*6:00 a.m.
*7:30 a.m.	6:15 a.m.
8:15 a.m.	7:00 a.m.
8:35 a.m.	8:15 a.m.
9:35 a.m.	8:50 a.m.
10:10 a.m.	9:30 a.m.
10:50 a.m.	10:10 a.m.
11:30 a.m.	10:50 a.m.
12:15 p.m.	11:30 a.m.
1:00 p.m.	12:10 p.m.
1:40 p.m.	12:50 p.m.
2:10 p.m.	1:30 p.m.
3:00 p.m.	2:00 p.m.
3:40 p.m.	2:35 p.m.
4:40 p.m.	3:00 p.m.
4:55 p.m.	3:40 p.m.
5:10 p.m.	4:25 p.m.
5:50 p.m.	5:15 p.m.
6:30 p.m.	6:00 p.m.
7:15 p.m.	6:30 p.m.
7:50 p.m.	7:10 p.m.
8:05 p.m.	7:45 p.m.
9:00 p.m.	8:40 p.m.
10:00 p.m.	9:40 p.m.
11:10 p.m.	10:20 p.m.
12:15 a.m.	11:30 p.m.
12:55 a.m.	12:30 p.m.
* Except Sunday	* Except Sunday

Black Ball Line

The BREMERTON SUN regularly published the Bremerton-Seattle ferry schedules. On the left is the schedule for June 1941; on the right the schedule for March 1943.

which could hold more than 200 workers replaced the cars' seats.

The Cannon Ball Express also took meals to the workers. Employees ate their lunches in old Northern Pacific passenger coaches, now given the names: The Feed Bag and The Ritz. These had been remodeled by the Joiner Shop. Lunches were prepared at the Supply Department Building and the two railroad cars were hauled to the piers during day and swing shifts.

Converted passengers cars transported hot food and beverages to workers on piers and dry docks. Tables and chairs replaced coach seats.

Puget Sound Naval Shipyard

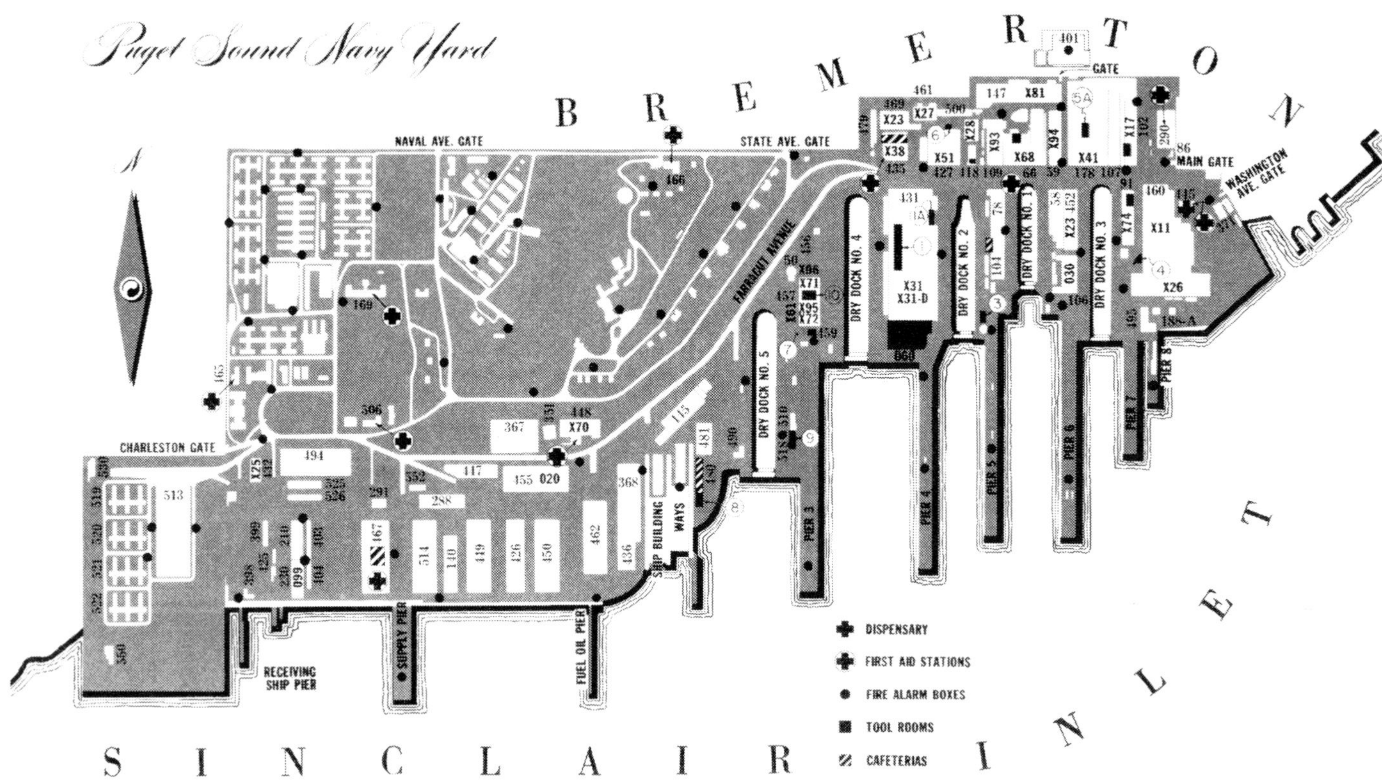

The Personnel Division compiled a "Directory of Bremerton Port Orchard Federal Housing Projects and Navy Yard Services". It listed important phone numbers and addresses and provided detailed maps, including the ones reproduced here and on page 111. Numbers preceded by letters X or O indicate the shops operating in the various buildings.

Puget Sound Naval Shipyard

The Feed Bag was given official recognition when Betty McGowan, in practice for christening the BREMERTON, cracked a bottle of lemon soda on a corner of the diner.

SALUTE reported:

There was a crowd, a gaily be-ribboned bottle, a mike system, maids of honor and a lot of laughter and enthusiasm as Betty swung heartily and sent the effervescent soda spattering in all directions in just one glorious try.

In addition, cafeterias were operated by the PSNY Cooperative Association.[50] During July 1944, workers ate 507,026 meals at an average cost of 29 cents per meal.

To ease the transportation problem, work hours and days off were staggered. Various arrangements were tried; seven days work with the eighth day off, 12 days work with Saturday off, 13 days with a Sunday off. Only two holidays were observed, Christmas and New Years. Half of each shop was off on one of those days.[51]

Mothers and wives hung blue-starred flags[52] in their windows and went to work in industry throughout the country, determined that sons and husbands in the war zone would have the equipment needed. Some followed husbands into the Yards. One of these was Gwen Masters[53], who started as a laborer on the badly damaged USS NEVADA, then was put in charge of other women hired to work on the ships.

Entire families worked in the Yard. There were the Shanleys, Tonges, Blanchards and Grandies, to name only a few. The contributions of the Harry C. Alguard family stretched from World War I to World War II and beyond. During World War I, Harry C. worked as a Joiner; his daughter Marie (Bobbie) was also employed in the Yard. She left during peacetime but returned to the Sheet Metal Shop during World War II. Her three sons followed her into the Yard.[54]

Flossie Tennison Jenson, whose husband served in the Navy during the war, recruited her family. Flossie hired on as a fry cook in August 1941; two years later she was promoted to cake baker in the main cafeteria, where she prepared 180 cakes a day. She brought her family from Missouri; mother Lenora Tennison, worked in Cafeteria 340, brothers Lee, Othel and Lyndon were in Shop 11, Shop 26 and the Naval Ammunition Depot, respectively.[55]

Seven Bidwells were newcomers from Scranton, Pennsylvania. Mother Bertha, three daughters, Mildred, Helen and Charlotte, and daughter-in-law Ann, worked in the Central Tool Room. Father Norman was employed as an electrician, and son Richard worked as an outside machinist. Two more Bidwell sons served in the Army.

Many Yeomanettes became Civil Service workers after World War I. At PSNY these included Lulu Brown, Maybelle Brown, Mae Christensen, Pearl Keas, Elizabeth Kelly, Velma McReynolds, Helen Miller, Jo Oass, Ethel Saunders, Lila Scheele, Catherine Ward, Agnes Williams, and Genevieve Wolfe.[56]

A three generation family of workers was made up of granddaughter Billie Hill and grandmother Marybelle Brown Craig (front row); son Harold Burton; sons-in-law Jesse Hill and Robert Angel (back row).

Centennial Book Committee

The rule to prohibit hiring more than one member of the same family or household was suspended for the duration of the war by the Civil Service Commission in July 1942. Four sons of Jack Shanley, a machinist working in Planning and Estimating, worked in the Yard. They were: Shipwright Ed, Yard Postal Clerk Jack, Boat Builder Raymond and Machinist Kiernan. A fifth brother, Jim, worked as Shipfitter at Mare Island Navy Yard.

Puget Sound Naval Shipyard

While most of the civilian women workers left after the war, some continued for many years at PSNY. Edith Cubbins, Gladys Duley, Catherine Hooper, Jennie Lowe, Florence Nichols, Eva Udey and Anna Wood were among them.[57]

Maude Campbell, who, with Lyle McLeod, became one of the Navy Yard's first female "detail draftsmen" during World War I, returned to the Design Section in 1943. She was one of the part-time employees recruited to help meet the work load. Fay Turpin, one of the rivet heaters and passers pictured on page 46 in Chapter 3, worked in Building 78.[58]

There were 80 women working in the industrial area in March 1942. By December 1944, 4,266 women performed jobs formerly done only by men; they worked as laborers, mechanics, chemists and draftsmen. William Filion,[59] who had been in the Director Shop since 1936, remembered the tremendous expansion of the Shop during World War II when:

> *. . . World War II brought a flood of women into the Navy Yard and I would guess about 200 eventually came into the Director Shop . . . they were put in the tool rooms and started at jobs easy to train them to do like sorting screws and repairing binoculars . . . Soon there were about 50 women in the Optical department alone — about equal to the number of men. Women also worked on repairing lead sights for small guns, until there were more women than men working on those. They were trained and adapted very well.*

A dress code and safety rules were issued for women; they were required to cover their hair and wear low heeled shoes. One piece, pocketless, navy blue denim suits were available for $3.49 by the fall of 1942; safety shoes soon followed.[60]

President Roosevelt authorized enlistment and commissioning of women in the Naval Reserve on July 30, 1942. Lieutenant Etta Belle Kitchen[61] was the first of the WAVES (Women Accepted for Voluntary Emergency Service) to be stationed at PSNY. The total number of WAVES at PSNY reached 250, almost the same as the number of Yeoman (F) during World War I. This time, 30 of the women were officers.

Integrating new workers into the work force had its problems; the turnover rate the first year was high. Because of the urgent need for employees, poor selections were sometimes made. For women, the attitude of male workers, unfamiliar noises and odors, strange machinery and lack of consideration for their special needs made adjustments especially difficult.

Representatives of women in military service during World War II, a Marine second lieutenant and two WAVE ensigns, pose in front of the Marine Barracks.

Centennial Book Committee

Two women operate a fork lift and a flat bed electric truck alongside Building 457. This 1945 photo shows historic Building 50 in its new location.

Puget Sound Naval Shipyard

Lorna Belden, gas attendant at the service corner of the Transportation Building, puts oil in one of the Yard's pickups. She was one of three women "grease monkeys" who tended the gas pump.

Centennial Book Committee

Hollywood star Lana Turner beams at the bearer of a flower tribute on June 16, 1942. The widely acclaimed sweater girl spoke briefly in support of War Bond sales. Nelle Klimas, secretary to the Captain of the Yard, served as Mistress of Ceremonies. Rear Admiral Taffinder stands at the right. Puget Sound Naval Shipyard

Many women were uncomfortable at home as well as at work, for the community was ill-prepared for the influx of women workers. Conditions improved when Muriel Job became Women's Co-ordinator in September 1942 and worked with the Production Department, city officials, USO and other community groups to improve the situation.

The workers put in long hours and needed recreation. In March 1944, Les Walters became director of newly organized Recreation Association.[62] Soon athletic teams, fishing derbies, pinochle tournaments and other diversions kept the off-duty workers occupied and happy. There were boxing matches, baseball tournaments and contests, including one for the best Christmas card verse. Bands played during lunch hours. The Puget Sound Navy Yard Revue, consisting of 30 talented performers and a 10-piece orchestra, presented novelty acts; vocalists and instrumentalists in 90-minute shows also toured housing projects in the county.

A public address system set up in 1944 featured a ten-circuit central control station and 100 watt amplifier at the broadcasting station, located near the Commandant's office in Building 78. Sound from eleven loud speakers on Farragut Avenue reached most of the Yard. The daily schedule included national news and reports of Yard production.

Workers had the opportunity to send recordings of their voices to folks back home. It was recommended that the message mention the Yard's urgent need for more workers. Except for restrictions on ship names, workers could say whatever they chose.

USO traveling shows brought top name performers: Bob Hope,[63] Ginger Rogers, Lana Turner, Jinx Falkenburg, Walter Pidgeon and others. Eleanor Roosevelt and high government officials visited the Navy Yard and Hospital.

Workers themselves performed at war bond rallies. One very popular singer was the Captain of the Yard's secretary, Nelle Klimas.[64]

The important job of getting out the Navy Yard news was augmented shortly before the advent of war. The first issue of the NAVY YARD SALUTE appeared on Friday, September 26, 1941, with an editorial staff of two: editor James Reems and illustrator James Hasting. Reems held the position until retiring in July 1973.[65]

SALUTE, begun as a four-page bi-weekly, soon appeared weekly with eight pages and more staff members. These included Jo Ann Oass, Jewel Linn, Don "Camera Queries" Allyn and Adele Ferguson.[66] Hank Ketcham, creator of Dennis the Menace, served a tour at PSNS, drawing cartoons for SALUTE as well as preparing promotional material.

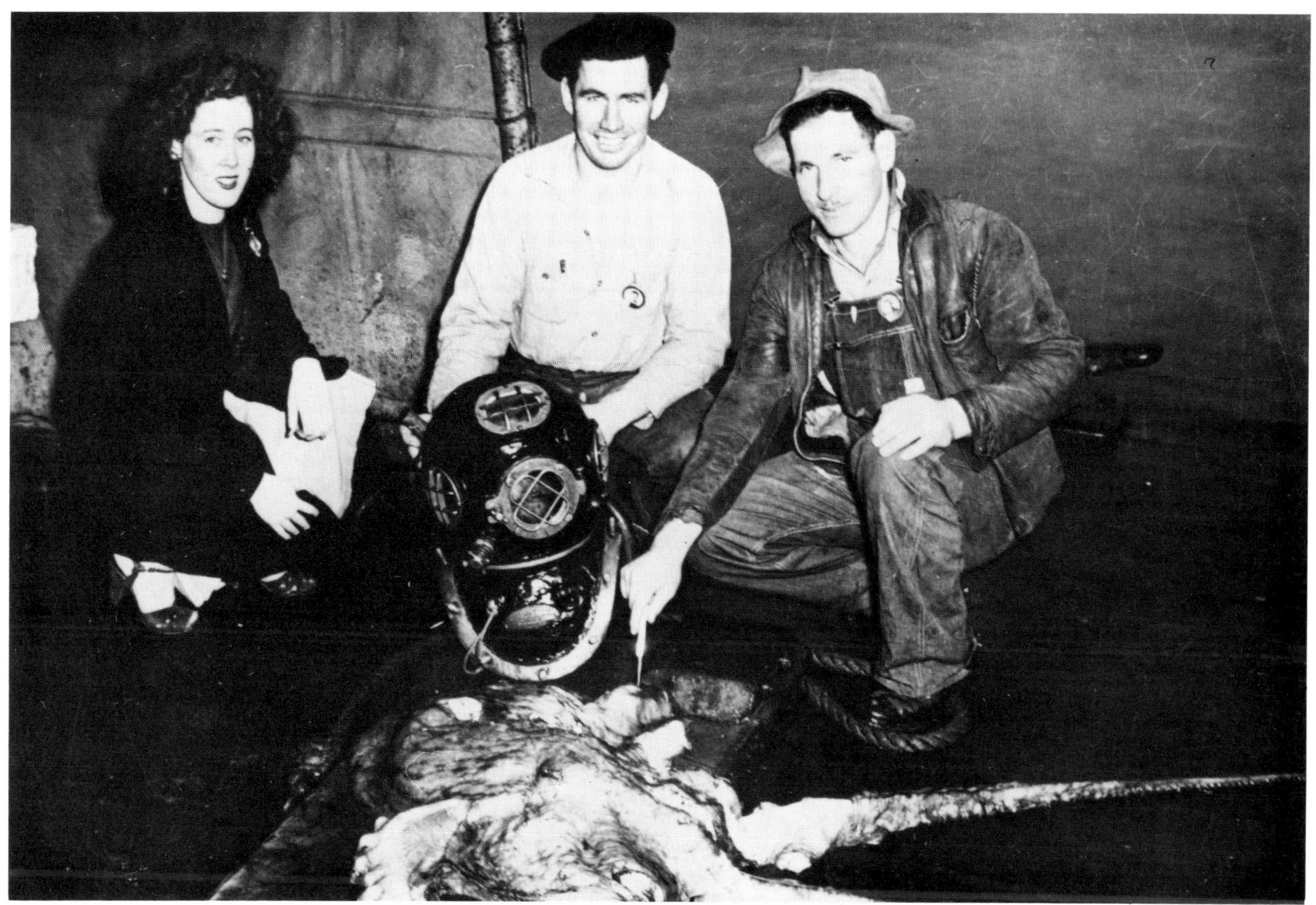

Diver Earl Lawrence, holding the diving helmet, shows SALUTE reporter Adele Ferguson his unusual catch. The octopus took Lawrence by surprise as he laid pipe off one of the piers. After a struggle, Lawrence slashed the creature and was finally able to struggle aboard the diving barge still encumbered with the wounded predator. Another deep-sea diver, Paul Leuthe, holds a screwdriver on one of the arms of the octopus. Puget Sound Naval Shipyard

Early issues pulsed with inspirational messages: Work hard, buy bonds, support the man in uniform, love our country. The paper also kept Yard personnel informed on new programs, goals anticipated, progress in other Yards and social and sporting events.

The war brought a new identification badge. The January 23, 1942, SALUTE reported on the new photographic type badges which eliminated possible borrowing of badges; 75-80 badges could be produced per hour.

Until shortly before the outbreak of the War, U.S. Marines had full responsibility for Yard security. When most of the Marine personnel left for the war zone, the Yard Police Department was formed. By November 1942, the organization had 103 members under Police Chief, Ensign Charles Lewis, formerly Bremerton's Chief of Police. A former police officer, W.E. Burnett, was Assistant Chief. Police assisted Marines in patroling the perimeters of the Yard 24 hours a day.

In the fall of 1942, the Navy Yard's fire department in the east end was supplemented by a second fire station, (Bldg. 503) in the west end. Fire Chief G. C. Triplett and Assistant Chief H.H. Henderson headed the complement of 28 civilian firefighters. The next year, an article written by Chief Triplett reassured the public of the safety of carbon dioxide fire extinguishers:

> *(carbon dioxide) . . . is to be respected but not feared. Do not become excited by the noise that is caused by the discharge of the gas from the horn, and, if in doubt, leave the compartment, as the smoke caused by a fire in a small, poorly ventilated space is more dangerous than the carbon dioxide.*[67]

Contests among the nation's Navy Yards and Stations encouraged safe industrial practices. PSNY was a consistent winner, taking first place in 1938 through 1944, and second place in 1945. This was in large measure due to the Industrial Department's investigation of industrial hazards and their consequent recommendations. Thanks to strict control of Navy galleys and Co-operative Association restaurants, food contamination and disease were kept to a minimum.

After the attack on Pearl Harbor, a large red cross was painted on the roof of the hospital; bomb screens were constructed around basement windows, auxiliary power for lighting was installed, and new buildings constructed. By the end of 1941, the hospital served the entire 13th Naval District,

Sick Officers Quarters had reached this stage of construction in September 1942. One of the former hospital buildings remaining in the Shipyard, it is now used for the Occupational Safety and Health Office for the Shipyard and Environmental Health Service Office for the Naval Hospital. Puget Sound Naval Shipyard

i.e., Washington, Oregon, Idaho, Montana and Alaska.

In addition to personnel stationed at the Yard, Receiving Barracks and Marine Barracks, the hospital cared for personnel of U.S. and foreign ships with Yard availabilities. Army personnel stationed in the area, Ammunition Depot and Torpedo Station employees with occupational injuries or disabilities and retired and Fleet Reserve personnel with their families swelled the numbers.

The average daily patient load was 742 in 1942. The peak daily load on August 21, 1944, was 1,041 patients. Hospital records show patients from 304 naval vessels were treated during the years 1941 through August 1945.

Yard personnel, particularly during the war, constantly developed new and better methods and equipment. Much of this was the result of the Beneficial Suggestion Program, commonly known as Benny Sugg. The program, which started in 1919, was completely revised in 1942. It was a great morale booster, and employees, from Helper to Master, sent in suggestions.[68]

Arvid Grenstad[69] and Robert Callison were awarded $1,000 for their work in removing film from optical lenses. Grenstad had been sent to the optical shop at the Naval Gun Factory in Washington, D.C., early in the war on a secret assignment involving coating of binocular lenses to increase light transmission at night. He and the other physicists returned to their Yards to set up optical lens shops: Grenstad wrote:

> *We moved into the new shop . . . in the summer of 1943 . . . It became apparent to me that in order to keep up with the workload a better means had to be found for removing old films that had been partially worn off . . . I went to work on this, and with the assistance of Mr. Callison, a chemist, worked out a method that in my view was satisfactory, and sent some samples to BUORD (Bureau of Ordinance) for approval. The method was not only approved but we received a letter of commendation, and a directive was dispatched to all navy yards to use our method . . .*

Grenstad's new shop was in the newly-enlarged Building 431. The Electric Shop had moved from Building 431 into 427 when Public Works moved its transportation equipment into newly constructed Building 455.[70]

The acquisition of land outside the Main Gate had been discussed for some time. With the outbreak of war, the need

All personnel working in the Director Shop on the third floor of Building 91 were required to have a Confidential clearance because much of the work dealt with classified material. Director Shop personnel cleaned, repaired, assembled and tested all types of optics for gun directors and searchlights. The Director Shop moved to Building 431 in the summer of 1943.
Puget Sound Naval Shipyard

was imperative. Assistant Secretary of the Navy James Forrestal wrote the Attorney General indicating it was now impractical to negotiate with the owners and recommending the land be acquired through condemnation procedures. On January 29, 1942, the Navy Department acquired from Sophia Bremer and others 0.586 acres[71] of land for $125,000.

The land was needed to relieve congestion inside the Main Gate. The planned expansion of the two-story Shipfitters Shop (Bldg. 460) would almost triple its size.

On March 29, 1943, more than 10,000 workers poured into the new building. The occasion was the official presentation of PSNY's first E Bond Pennant and to allow as many workers as possible to see the new plant. Its 347,000 square feet of working space was designed to meet wartime demands. By mid-1945 over 6,000 shipfitters, welders, helpers and mechanic learners were involved in the fabrication of all structural members of ship hulls.

The newly enlarged building covered the former site of Building 157. The Supply Department had accomplished its move from that structure into its new seven-story Building 467 in the west end of the Yard in record breaking time.[72]

Something less than 24 acres of land was purchased outside the southwest boundary of the Yard in 1942 for $431,421. Buildings on the land were destroyed or moved. The Navy Yard constructed a quay wall as a continuation of one in the Yard and relocated the railroad car-barge landing to that area.

A little more than half an acre of land near Warren Avenue and Gregory Way[73] had been purchased in 1938 for $10,000 in order to extend the southwest corner of the Foundry. This increased its floor space by 60 percent. John Sendner was the master of the shop, which became one of the finest in the Northwest.[74]

As a result of the increased need for explosives, the Ammunition Depot on Ostrich bay was taxed to its limits. Captain F. P. Wenker, who had been Executive Officer there, wrote, in 1982:

Since NAD Bremerton could not be safely or economically expanded and the necessity for more storage space became apparent in 1941, a large bomb and mine storage facility was built on Indian Island, across

Buildings outside the Main Gate were purchased early in World War II. The YMCA is in the left background. To the right of the "Y" are two buildings, on land purchased during World War I, and the new Industrial Dispensary.

Puget Sound Naval Shipyard

During the Yard's expansion to the west in 1941, the American Legion Hall was moved across to the west side of Cambrian Avenue. The only other building moved and still existing is the White Pig Tavern on Callow Avenue, still in operation.

Puget Sound Naval Shipyard

an estuary from Port Townsend . . . After the start of the war, further expansion was required and the area now known as NAD Bangor was acquired . . . After this building was under way the Bureau of Ordnance initiated the building of the railroad from the nearest railhead to Bangor, NAD Bremerton and the Navy Yard, thereby eliminating the car barges.[75]

Bremerton was listed in Ripley's BELIEVE IT OR NOT columns as the largest city in the country without direct rail connection. Although a railroad had been expected early in the century, it did not come until World War II. Construction, under supervision of the Bureau of Yards and Docks, started early in 1944 on the government line from Shelton to Bremerton and Bangor. It was turned over to the Northern Pacific Railroad for operation on April 10, 1945. The first ammunition train, consisting of 33 freight cars, arrived at Bangor on April 14, 1945.

The first car of freight reached the Navy Yard on June 5; on Tuesday, June 19, Bremerton and the Navy Yard celebrated the arrival of the first passenger train. This was not the beginning of regular service, but a special three-car inspection train marking the official opening of the branch railroad, "properly loaded with Navy gold braid, mayors and other civil bigwigs of the principal cities of the Olympic Peninsula."[76]

After the arrival of the railroad, the Navy Yard replaced its two steam locomotives with two diesels. George Drew Johnson, conductor on one shift, started work in the Navy Yard in April 1942:

> *All materials came to the shipyard by barge or truck. The barges were railroad car barges . . . We used the large engine for taking the cars off the barges, sometimes three or four times a shift, and distributed them to various places. The small engine was used to shift the cars around the dry docks and into buildings where they were needed . . . then we reloaded the barges with empties . . .* [After the railroad came] *we sometimes had to go to NAD and Bangor.*[77]

One of Navy's staunchest supporters died April 12, 1945. Franklin Delano Roosevelt had supported the Navy from 1914 as Assistant Secretary of the Navy until his death while President of the United States.

The President's September 1942 visit to the Puget Sound Navy Yard was kept a secret, even though he spoke to Yard workers. According to Nard Jones[78]

Conductor George D. Johnson, left front, poses with Engineer Donald St. Peter, Brakeman George Hoffman and two railroad men at Bangor Naval Ammunition Depot. George D. Johnson

That mysterious visit of President Roosevelt to Puget Sound was the best kept secret of the war except for the top-security business at the plant over the mountains in Hanford, Washington.

Roosevelt visited the Navy Yard again in 1944, at which time he broadcast a nationwide talk from the prow of destroyer USS CUMMINGS (DD 365)[79] in Dry Dock 2. He praised the Yard for its achievements, making special mention of the repairs and return to the fleet of the battleships damaged at Pearl Harbor.

Less than a month after FDR's death, war in Europe ended with the surrender of Germany on May 8, 1945. To a generation that had grown up believing that Germany's militaristic/expansionary aims were caused by the signing of an Armistice in 1918 instead of a surrender, nothing but total surrender was acceptable this time.

Troops in Europe awaited their transfer to the western arena where the war still raged furiously. The siege of Okinawa continued into its second month; Japanese suicide planes still hammered the Fleet.

In June when organized resistance collapsed on Okinawa, the last island barrier to the Japanese mainland was gone. Both sides increased preparations for the expected invasion. At Japanese ports, massive fortifications awaited the assault; Navy ships swept mines from the Yellow Sea.

Then on August 6, an atomic bomb was dropped on Hiroshima, Japan, by the United States, the first country successful in the attempt to build a nuclear fission bomb. On August 9 a second bomb fell on Nagasaki. On August 14, the Japanese government surrendered.

Once again, the people of the Puget Sound Navy Yard learned the big news in different ways. Arvid Grenstad was at work in the Director Shop - and continued to work. Ralph

President Franklin D. Roosevelt arrived at Puget Sound Navy Yard aboard USS CUMMINGS (DD-365) on August 12, 1944. From a podium in front of the number one gun mount, the President delivered a speech that was broadcast nationwide. Larry J. Jacobson

Three women in the Sheet Metal Shop receive word of the surrender of German forces in Europe. Their mixed emotions range from relief to concern about those service men still fighting in the Pacific. These three, Jane Miles, Laura Steele and Pearl Wilkes were the mothers of seven servicemen. Puget Sound Naval Shipyard

VJ Day crowds in front of Forget-Me-Not Florists and Olberg Drugs at Pacific and Fourth, force many pedestrians into the street. At 4:04 p.m., August 14, 1945, Frank Lyons read the Navy Yard Commandant's announcement that the war was over. Fire station sirens wailed on downtown streets, shredded papers fluttered from third story windows.

Puget Sound Naval Shipyard

Bob Crial, Industrial Relations, stands beside Building 78/104 to display long-awaited news. The subheadline reads, TRUMAN ANNOUNCES JAPS HAVE ACCEPTED TERMS. Hope Beck

The small ferries to Port Orchard and Annapolis faced their biggest challenge on VJ Day, August 14, 1945. The waiting line zigzagged back and forth on sidewalks and streets as far as the Main Gate. At the end of three hours, all passengers had been delivered safely to the other side in eleven boats called into service. Mary Lieseke, Horluck Ferries

Smith had just come in to start the swing shift when his leading man told him to go home. The war was over.

Smith says:[80]

It was the biggest traffic jam you've ever seen, with both shifts leaving at the same time. The bridge was wall to wall people and cars. I walked home to Sheridan Park, along with many others.

Parker Snapp recalled:

At war's end, the Shipyard emptied in a shouting flood . . . in time to board the Kalakala for return to Seattle. An apprentice was boosted by his companions to the canopy over the fantail. He crept foward, cut the airhorn line, tied a cord to the stub and brought it back to the stern where it was pulled taut and tied to a stanchion. The airhorn blasted for three minutes or more before the skipper and crew identified the problem.

According to Jerry Grosso:

. . . in Port Orchard, Bay Street was crowded with people whooping and cheering. Some emptied their pockets and small change clattered off store windows. All whistles on ships that could, were blown. In the Shipyard, thousands of workers put down their tools and went off to celebrate, [some] *not even coming back to pick up their pay.*

However, devotion to duty did not end with the ringing of bells and blaring of whistles. In the Accounting Department, reliable and trustworthy employees remained to get out the payroll.[81]

The war was over. It was time to celebrate.

Ship's company of USS LAERTES (AR 20), newly arrived from overseas duty in the Marshall Islands, held a dance at the American Legion Hall in December 1945. Smiling Mary and Phil Drouin, left center, and others celebrate their homecoming. Philip Drouin

An impressive ceremony took place on USS MISSOURI when 80 foreign-born men and women became citizens of the United States. The naturalized citizens raised their hands while standing on the World War II Surrender Deck on Veterans Day, November 11, 1954. National press, radio and TV covered the event; commemorative envelopes honored the occasion.

Kitsap County Historical Society

Chapter 6

Peace and a Police Action

1945-1955

Suddenly the Yard had a new mission: "Magic Carpet", an operation in which warships would be converted into transports to bring troops home from the Pacific war zone. Shipfitters and welders set a record in November 1945, installing 3,000 bunks in USS BUNKER HILL (CV 17) in only 16 working hours.[1] Her hangar deck covered from bow to stern with pipe and canvas bunks four high, the aircraft carrier could accommodate 4,500 returning service personnel.

At the Yard's west end, the Receiving Barracks became the Separation Center where returning servicemen poured in when they had enough service points to be demobilized. Points were accumulated through length of service, time overseas and combat campaigns.

The fleet was cut back. Some ships were scheduled for inactivation, others sold for scrap. Almost 100 other ships were involved in Operation Crossroads, the Bikini Atoll atomic bomb tests, and never used again. Sadness was expressed at this fate for ships like SARATOGA, PENNSYLVANIA and NEVADA, major contributors to the Navy Yard's workload throughout the years.

Although many workers left voluntarily at the end of the war, the smaller work load spawned a new acronym, "RIF", meaning Reduction In Force. Veterans received five points for their military service, ten if wounded or disabled, which gave them preferential status for retention.

The first big layoff came in January 1946, with 970 journeymen and helpers discharged. Three months later another 2,880 workers were dropped from the Yard payroll. The decline continued. By the end of 1946, there were less than 9,000 employees, who worked mainly on ship inactivations and routine overhauls.

To ensure a stable work load for the Yard and retain as many skilled workers as possible, the Puget Sound Naval Base Association was organized in April 1946. Previously the Bremerton Chamber of Commerce and the Bremerton Metal Trades Council had maintained committees to promote the Sinclair Inlet area's general economic welfare. Now these efforts were combined.

The new association sent its first president, Stan Oliver, to recruit Rear Admiral Claude S. Gillette, USN (Ret),[2] as the Association's representative in Washington, D.C., where he would maintain liaison with Washington State members of Congress.

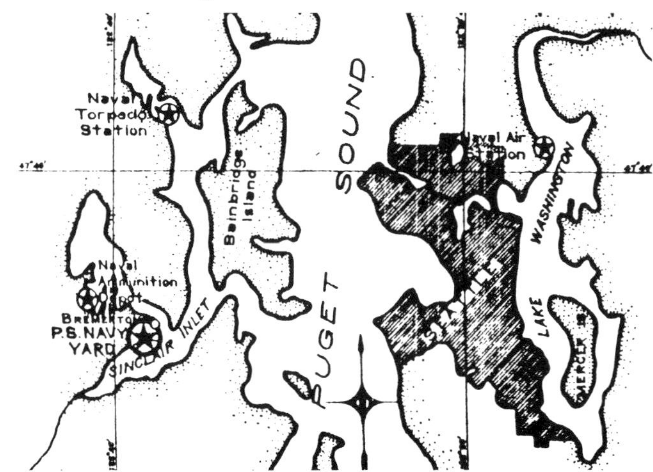

The Thirteenth Naval District in 1940 included Puget Sound Navy Yard and Naval Ammunition Depot at Bremerton, Naval Torpedo Station at Keyport and Naval Air Station at Seattle. Puget Sound Naval Shipyard

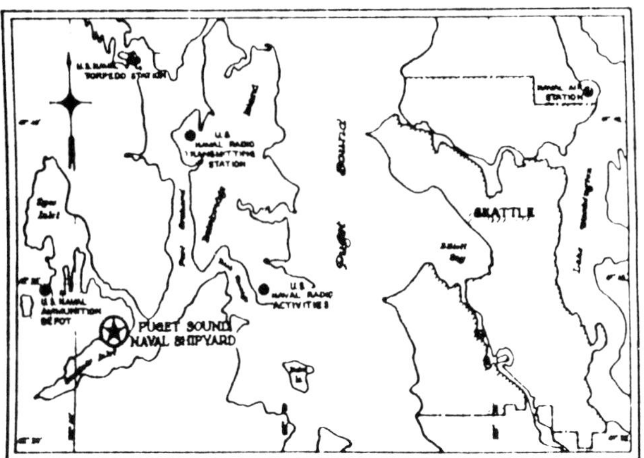

The roster of the Thirteenth Naval District activities in 1945 added two Naval Radio Stations on Bainbridge Island and a renamed Puget Sound Naval Shipyard.

Puget Sound Naval Shipyard

New additions by 1951 included Bangor Annex NAD, Degaussing Range and the Naval Fuel Annex at Manchester.

Puget Sound Naval Shipyard

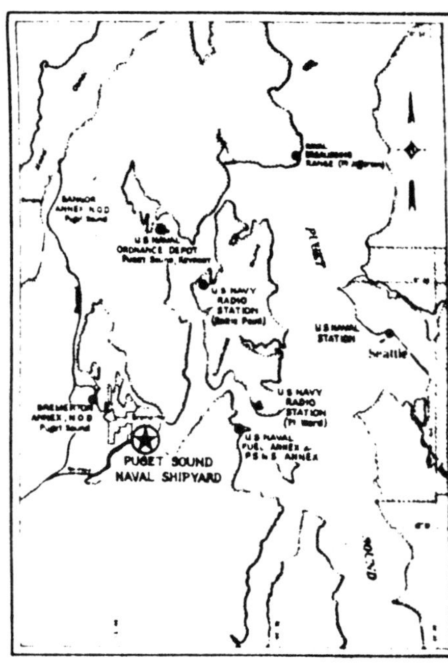

Late in 1945, the Puget Sound Navy Yard received a new name, Puget Sound Naval Shipyard. As such, it was part of the new Puget Sound Naval Base. Rear Admiral Ralph W. Christie[3] became Commandant of the Naval Base, which included the Shipyard, Marine Barracks, Naval Hospital, Naval Barracks (former Receiving Station), Inactive Fleet Berthing Center, Naval Ammunition Depot, Bangor Naval Magazine, Keyport Torpedo Station, and Indian Island Magazine.[4]

Commodore Webster M. Thompson[5] became Commander of the Shipyard, which was reorganized into seven departments and two staff divisions. The office of Captain of the Yard was abolished and Captain Willliam B. Coleman became head of the Administration Department. The other departments were Planning, Production, Public Works, Supply, Medical and Fiscal. The two staff divisions were the Industrial Relations Office (IRO) and Management Planning and Review (MP&R).

In 1947 the Inactive Fleet Berthing Area became a separate command: the Bremerton Group, Pacific Reserve Fleet. Foreseeing the need, a site for berthing inactive ships had been discussed during the war. One suggestion centered on damming and locking Dyes Inlet to provide space for anchorage. The Bureau of Ships proposed initially that the facility be built near Port Orchard but acceded to Commodore Thompson's recommendation to berth the inactive ships outboard of tidelands southwest of the Yard, adjacent to the government's new railroad right-of-way. New mooring facilities were completed in November 1947.[6]

The mooring piers for the Inactive Fleet berthed a variety of ships, including the former luxury ship SS GREYHOUND (IX 106), carrying two stacks; she had been used as a Barracks Ship during World War II. Others in this 1946 photo are, front to back: USS MARYLAND (BB 46), USS SABEATA (YTV 287), USS INDIANA (BB 59), USS ESSEX (CV 9) and USS BUNKER HILL (CV 17).

Puget Sound Naval Shipyard

Five Masters received Navy Meritorious Service Awards late in January 1947 for their "outstanding contributions to the Navy Yard work throughout the war years". The Masters are, left to right, front row: Theodore Peterson, Machine Shop (Inside); Herman Petersen, Shipfitter; and William Wesseler, Electric; back row: Russell Hibbard, Sheetmetal; and Walter Bruns, Riggers and Laborers.

Puget Sound Naval Shipyard

Among those who received Meritorious Civil Service Awards[7] for outstanding contributions to navy yard work during the war years, were the masters of five Shipyard shops. They were Herman Petersen, Russell J. Hibbard, Theodore Peterson, Walter Bruns and William Wesseler.

Herman Petersen[8] was the dean of the masters, having entered the Navy Yard in 1902 as a "boy" earning 78 cents a day. He became master of the Shipfitters Shop in December 1919 and masterminded its huge expansion during the late 1930s and early 1940s. He retired in the summer of 1948.

"Theo" Peterson[9] retired in 1951. Theo entered the Yard as an apprentice in 1906, left in 1911 to major in mechanical engineering at Washington State College, returning to the Yard in 1915. He advanced to leadingman as soon as he reached the required age of 25. He became master of the Inside Machine Shop in 1936 and directed the additions to Building 431 which made it the largest machine shop in the United States.

When Russell J. Hibbard[10] retired in 1953, there had been a Hibbard as master of the Sheetmetal workers for all

26340 M. D. Currey

PUGET SOUND NAVY YARD

EMPLOYEES MUSTER CARD

SEX ____ Ht. ____ Age ____

Color ____ Wt. ____ Date 3-16-45

The Muster Card, an aid in keeping attendance records, came into use for only a short time at the end of World War II. Workers handed their cards to their leadingmen who noted attendance in their notebooks, and returned the cards at day's end. After a year or two, according to Currey, the Muster Card was discontinued.

Monty D. Currey

The eastern third of the 400-foot long Shipfitters Shop (Building 460) constructed in 1940, shows in the upper left of this picture. The 1943 expansion of the building to the north more than tripled its size. The Industrial Clinic (Building 445) stands at the left.
Puget Sound Naval Shipyard

The Shipyard's contribution to Seattle's Parade of Progress included a long display presented by various shops. Clearly visible are photos, equipment and parts used in the work of the Metallurgical Lab, Foundry, Supply Department and Machine Shop. Farther down the line are two model plastic boats and a laminated boat section.
Kitsap County Historical Society

A big attraction at Seattle's Parade of Progress was underwater welding demonstrations. The square structure on the right is the glass water tank in which divers worked with welding and burning equipment adapted for underwater use. Note diving helmet on tank shelf and posted diving schedule. Kitsap County Historical Society

but two of the previous forty years. His father, Russell E., was master from December 1911 to February 1934. The next two years other masters double-hatted as master of Shop 17, Shop 11's Herman Peterson in 1935 and Shop 41's George Penketh[11] in 1936. R. J. Hibbard was master of Shop 17 for the next 17 years.

More than 20 other masters retired before the end of 1954, but Bruns of the Riggers and Laborers Shop and Wesseler of the Electricians, masters only since 1944, remained in the Yard into the 1960s. Wesseler was active in planning the Shipyard's exhibit at the International Labor Organizations' Pacific Northwest Parade of Progress Exposition held at the Seattle Field Artillery Armory in June 1946.

Also involved in planning were Master Patternmaker Edwin S. McAfee,[12] Machinist Foreman Fred K. Hicks[13] and Shipyard Co-ordinator Captain Raymond O. Burzynski.[14] BREMERTON SUN editor Julius Gius and Bremerton Chamber of Commerce Secretary Norman Box coordinated the city's participation.

Puget Sound Naval Shipyard's display dominated the Navy's allotted 2,300 square feet. The display included a demonstration of the "magic weapon of World War II" — radar. Visitors, looking into a video screen, saw ship movements in Elliott Bay. A scale model[15] of the Shipyard and transparent plexi-glass models of an aircraft carrier and an assault transport also intrigued visitors.

Another big attraction was the Shipyard's demonstration of underwater welding, in which divers wearing "hard-hat" suits worked in a 5,000-gallon glass water tank and explained their work over a loudspeaker. The first truly successful under-water welding operation took place in the waters off the Puget Sound Naval Shipyard. Welding and burning were an integral part of the diving school curriculum. Diver Earl Lawrence explained:

> *Because of PSNY's good facilities and the willingness of the various shops to support the divers by building new equipment, we were able to improve on all underwater work and carry the divers to greater accomplishments under water.*[16]

Lawrence and his divers[17] developed an underwater burning torch and technique for cutting heavy steel under water. These were made available to civilian divers in 1947 when the DIAMOND KNOT, a commercial ship carrying 254,000 cases of canned Alaska salmon, sank after colliding with FENN VICTORY in the Straits of Juan De Fuca, 15 miles west of Port Angeles. However, the commercial divers, unfamiliar with the technique, were not able to cut the shell plating on the side of the ship to reach the cases. Lawrence and another PSNS diver[18] took leave and completed the job in two days.

To prevent electric shock and minimize electrolytic action[19], Lawrence devised a method to apply rubber coating on hard-hat diving helmets. This technique was then adapted for use on other metallic objects used under water.

The Yard's rubber division had been established in the Inside Machine Shop under Supervisors Bernard J. Oliver and Fred K. Hicks during World War II. Rubber coating of ships' parts could then be done in the Navy Yard, and workers were able to control the quality of rubber used to coat propellor shafts as protection against corrosion and to manufacture rubber and plastic parts that were in scarce supply.

When Wah Bong Lew[20] came from Mare Island in 1946 to provide expert rubber technology, he developed new sealing techniques to weatherproof decommissioned ships. Lew developed polysulfide compounds for caulking wooden flight decks on carriers and patented an underwater rubber boot seal process. The latter was used to keep sodium chromate solution from leaking out of the boot around the rubber bearing staves for corrosion prevention on inactivated ships.

He also developed Navy Rubber Formula 112 which sealed the domes put over gun mounts on inactivated ships to maintain low humidity and prevent rust. Machinists Paul A. Kiefert and Lawrence Lundgren did the necessary shop experiments with Lew's formulas, which were later adopted by the commercial rubber industry.

Two new shops were established in the Yard in 1948, Electronics and Chronometer. Before 1948 all manufacture and repair of high precision Navy chronometers had been done at the Naval Observatory in Washington, D.C. Richard Linkletter of the Shop Superintendent's Office met with naval and civilian timing experts and developed plans for a new shop at the Puget Sound Naval Shipyard. Leadingman Louis Lancaster transferred to PSNS to head that shop.

The Chronometer Shop's highly trained technicians were part of the Director Shop[21] and thus under the cognizance of Theo Peterson. A complete overhaul, required every three years on all Navy chronometers, provided a continuing source of work for the Yard.[22]

Foreman Thomas Hall[23] headed new Shop 67, Electronics. Hall had been in the Radio Material Office, repairing all types of communication and navigation equipment in Building 466, which John Mason[24] remembered had been "especially constructed" on the hill above the officers' quarters:

. . . away from the industrial area to minimize electrical noise. Also, all reinforcing steel rods in the structure were welded at intersections to avoid the generation of electrical noise.

Diver Earl Lawrence displays old hardhat diving helmet. This was the original helmet that was rubber-coated at PSNS. Richard Linkletter

The civilian head of RMO was radio communications pioneer Charles E. Williams, whose principal assistants were supervisors H. Douglas Finch and Arthur H. Brudwig for shore stations. Fred N. Gruwell for ships and Roy Levin for drafting.[25] After the Electronics Shop was established, the shore station activities of the RMO were transferred to Seattle; shipboard electronics activities were handled in the Shipyard Design Division.

By the end of the 1940s, more than half of Shop 67 employees worked under Quarterman L. P. "Pat" Campbell[26] on a classified project later known as SERAD, Special Electronics Restoration and Distribution.

Because new electronic equipment was in short supply and expensive, the shop repaired and upgraded inoperable

Bremerton Mayor Lineham "Hum" Kean greets President Harry Truman and Governor Monrad C. Wallgren at Gorst in June 1948 before driving with them to Bremerton and the Puget Sound Naval Shipyard. Ruth Kean Yates

electronic equipment returned from forward areas of the Pacific. The Central SERAD Coordinator in Washington, D.C., said every dollar spent on the project put $10 worth of equipment into the Navy Supply System. This SERAD experience also trained mechanics in the rapidly expanding field of electronics.

President Harry S Truman visited Bremerton and the Puget Sound Naval Shipyard in 1948. Before entering the Shipyard, the President addressed a crowd in front of the Elks Temple at Pacific Avenue and Fifth Street. Someone called out, "Give 'em hell, Harry!," a phrase that figured prominently in his election campaign.

Crowds lined Farragut Avenue shortly before noon on June 10 as the car bearing the President entered the main gate. After a stop at Building 78, review of the Marine Detachment and a tour of the hospital grounds, the President boarded Washington State yacht OLYMPUS and sailed for Seattle.[27]

The Shipyard was involved briefly in the ferry business in 1948. During a strike of the Black Ball Ferry Company, LST 1085 sailed to and from Seattle with 650 straight-back chairs on its cargo deck for the 600 to 900 workers who commuted daily. It was cold and uncomfortable riding, but it served the purpose — and it was free.

The Shipyard's Eye Correction-Eye Protection Program to detect workers who needed corrective eye glasses began that September. Safety glasses previously had been plain

Russell Hegdahl at his drawing board. As Shipyard Illustrator he prepared art-work for the SALUTE, training aids, billboards and artists conceptions of ships and buildings. Mrs. Russell Hegdahl

Upon pumping down Dry Dock 3 in January 1949, the largest haul of herring in the memory of shipyard workers was caught within the dock. This was not the first or last time large numbers of fish were caught in any of the shipyard's dry docks; however, over the years effective fish deterrent methods have been developed and fish captures are minimal.

Puget Sound Naval Shipyard

glass or a cover for the employee's personal glasses. Now employees obtained safety glasses with their own prescriptions. The year 1948 also brought the start of required annual X-ray exams intended primarily to discover tuberculosis but which also pointed out other medical problems.[28]

Puget Sound Naval Shipyard won the Navy's top Safety Award in 1948. As part of its Safety program, the Shipyard published the RIGGERS' HANDBOOK AND SAFETY MANUAL in 1949. The pocket-sized book, full of information on the right way to do rigger jobs, was compiled and edited by Safety Superintendent K. I. Twitchell and illustrated by Russ Hegdahl.[29] Others involved were Walt Bruns, George M. Hill, Fred Timmerman, Earl R. King, and Clarence Bills.

In the spring of 1949, shipyard personnel began the task of constructing and placing into operation a high-powered Navy radio transmitting station in Jim Creek Valley between Wheeler and Blue Mountains, 55 miles northeast of Seattle. The 1,000,000-watt transmitter became operational on November 18, 1953.[30]

When pumping down Dry Dock 3 after an undocking in January 1949, the Shipyard discovered the dock was full of herring — an estimated 700 tons of fish. Pumping was stopped, the dock reflooded and the caisson opened. Shops 11 and 72 rigged three paint scaling guns on steel plates and lowered them into the water from three boats which swept down the dock from the head end. The guns' triggers were tied down so that a steady tattoo against the plates would drive the fish from the dry dock into the Inlet. The noisy time-consuming operation was not effective and was not repeated; better methods of keeping fish from entering the open docks were devised[31].

Captain William E. Sullivan[32] became Commander of the Puget Sound Naval Shipyard on January 25, 1950. He

Loft riggers on the third floor of Building 91 manufactured landing nets used by troops when disembarking from ships. These particular nets, made in 1950-51, served Marines and Army personnel when landing in Korea where extremely high and low tides made piers impractical. The nets were made of 2¼ inch manila line. Riggers seen here are, left to right: Sailmaker William Dollard, three unidentified men, Loft Rigger Jack Hines and Sailmaker Harold O. Pickard.

Harold Pickard

had been selected for promotion to Rear Admiral but during his promotion physical was discovered to have cancer. He died shortly thereafter. On May 7 a memorial service was held for him at the Naval Base chapel before his body was taken to Arlington National Cemetery for a military funeral.

This was the third time the Yard buried its senior officer. The others were Rear Admiral Ziegemeier in 1928 and Rear Admiral John Halligan[33] in 1934. Halligan, too, had been in charge only briefly.

Rear Admiral Hugh E. Haven became the new Commander on May 17, 1950. The following month Puget Sound Naval Shipyard was again on a war-time footing as the Korean "police action" began. Haven summed up the feelings of many:

To Shipyard workers who were here during the late 30's when the nation was frantically arming for defense and who served on the production front during World War II, the course of international events has an ominous ring of familiarity . . . For us all there is a queasy feeling of this is where I came in. A depressing feeling of having to go through the awful mess all over again . . . We know that although no war has been declared, we may be hovering on the borderline of World War III.[34]

By August, 16 ship activations were under way; the ships that had been "mothballed" such a short time before were back in the dry docks. Most of these were landing ships or smaller, but six small "jeep" carriers (CVE's) as well as three larger carriers were sent to Korea. PRINCETON (CV 37) was the first of the larger ships to leave.

USS ESSEX (CV 9) and USS BON HOMME RICHARD (CV 34) were re-commissioned in a dual ceremony on January 15, 1951, with Fleet Admiral Chester A. Nimitz as principal speaker. Both he and Haven lauded the shipyard workers for completing ESSEX in two-thirds of the scheduled time.

By December 1950, USS BRUSH (DD 745) and MANSFIELD (DD 728) were in the Yard for extensive hull damage suffered from mines in Korean waters. Although U. S. Navy ships were actively engaged in Korean operations, the Shipyard was not as heavily involved in battle-damaged ship repair during this period as during World War II.

The work force increased from 7,800 in mid-1950 to 15,300 in June 1952, followed by a gradual decline in employment. United Nations and Communist negotiators signed an Armistice agreement in July 1953. Negotiations continued, but the threat of a full-blown war was set aside.[35]

Royalty, in the person of Emperor Haile Selassie of Ethiopia, arrived at Puget Sound in June 1954. At PSNS, he inspected the Marine Honor Guard, toured the waterfront, observed the activities of the Boat Shop, Inside Machine and Foundry Shops, and enjoyed a baked salmon dinner at the Officers' Club. The Emperor's visit was initiated by his desire to view the plastic boats built in the Yard.

Boat building was an important part of the Yard work load. During the 40s and 50s the shipyard was recognized as the leader in the field of laminated wood and resin plastic use for both naval and civilian use. Charles V. McLaughlin, master of the Boat and Joiner Shop since 1927, was a leader in the development of the shop's modern boat building techniques.[36]

Following World War II, when supplies of large size ship construction grade oak were depleted, the shop, under the direction of Chief Quarterman Donald May, developed the process of laminating wood. By 1950, PSNS turned out plywood-sided, laminate-framed 40-foot personnel boats. PSNS was the only shipyard using laminated oak for structural members.

Two wooden mine sweepers, built for the Dutch government in 1953, were the first new-ship construction work in the Yard since the war years. The shop fabricated the frames of AMS-167 from white oak, which was laminated straight and then steam-bent to shape. The AMS-168 was the first

Six hefty men test a laminated staging plank in the Boat and Joiner Shop in November 1948. The Douglas fir plank, measuring approximately 2 inches x 12 inches x 12 feet, supports the total weight of 1,258 pounds of (left to right) Art Spute, Fred Olson, Ernie Johnson, John Morgan, Art Davis and "Red" Jamison. John Morgan

Navy-built vessel in which fir frames were laminated to shape.

Some boats were already being built of plastic, but construction by the open mold method was slow and complicated. In 1946 Engineer W. H. "Bill" Miles,[37] then of the Scientific and Test Group, and Metallurgist John Tenold began experiments impregnating canvas with a new polyester resin made by the Bakelite Corporation. When they were successful using this to coat balsa life rafts, they developed a small boat model. The Bureau of Ships was interested and in 1948 allocated $50,000 for the construction of plastic 12-foot wherries.

Shop Superintendent, Commander Richard C. Lombardi, set up "Operation Plastic Boat" in Building 58. Fred Mills, head of the Material Laboratory Branch, organized the group which included sailmakers, machinists and patternmakers. First Bill Miles and later Chemical Engineers Earl Thompson and John Turbitt directed the laboratory crew in experiments with the new type of polyester resin. Boat builders, pipe coverers and insulators joined the group[38].

According to SALUTE, the breakthrough for mass production came with the decision "to switch from the use of molds for matt forming to that of using plastic jigs, thereby freeing the molds for turning out additional boats."[39] The Shipyard produced more than 80 wherries which were strong, durable and required little maintenance.

Early in 1952, the Shipyard introduced its 36-foot plastic landing craft, the largest marine plastic molding in the country. Tested on Blake Island, in Arctic waters and in more than 40 beach landings in California, the boats proved unsinkable.

At the end of 1953, Central Tool Shop Foreman Gordon M. Erickson[40] set his crews to develop the use of epoxy plastics for the manufacture of tools of all sorts. Within a few years, Shop 06 was producing lightweight tools for short-run operations in most Production Department shops. The Material Lab and Woodworking Shop began testing the use of plastic coatings on ships' hulls and rudders for protection from salt water damage.

A major change took place in July 1953 when the Fiscal Department became the Comptroller Department, Code

Several years before the boatbuilders constructed plastic boats, they built laminated plywood boats. This December 1944 photo shows the laminated keel of a 50-foot motor launch.

Puget Sound Naval Shipyard

The Accounting Department in mid 1940's. Front row, left to right: R.F.M. Rotsaert, Captain M.E. McMahon, Ensign Cooper and Genevieve Wolfe. Back row: Susan Wilcox, Mrs. Reed (in charge of Bonds), Mrs. Peterson (Disbursing), Andy Billings, Max Josephson, Pearl Keas, T.F. "Bill" Drake and Connie Osburn. Dorothy Germain

600. There were four major divisions: 610, Internal Review under Max S. Josephson; 620, Budget and Statistics, transferred from MPR with W. S. "Bill" Josephson still in charge; 630, Accounting and Disbursing under former Yeoman (F) Genevieve Wolfe; and 660, Disbursing, headed by Gerald V. Stoffel. Captain Marvin Gluntz, an engineering duty officer, was the first Comptroller.[41]

Code 610 was tasked with planning for conversion to the Navy Industrial Fund. Max Josephson explained:

> *The installation of the Navy Industrial Fund financing essentially put the Shipyard on a business basis comparable to private industry. Prior to this time, the Shipyard received direct appropriation funding for work. Under the Industrial Fund, we received an allocation of funds as working capital, our inventories were capitalized and a complete set of accounting reports . . . were installed.*[42]

All work performed, including overhead costs, had to be paid out of working capital prior to billing. Labor, material and overhead costs were billed to the customer, whose payments replaced the funds expended from working capital.

Naval Shipyards were the first facilities to operate on the new system. On July 1, 1954, Puget Sound Naval Shipyard began operation with a capital fund of $16,100,000, plus $900,000 worth of materials already in inventory in the Yard.

By 1954 the Yard again was involved in the inactivation of ships, the most memorable being that of USS MISSOURI (BB 63) because of her historic significance and size. Work done on the 887-foot long ship in 867-foot long Dry Dock 2, where her bow extended over the head of the dock, was referred to as "Operation Shoehorn".

National holidays had always been celebrated with ceremony and fervor in Kitsap County, especially Memorial Day and Armistice Day. One of the most impressive occasions occurred in 1954, the first year after Congress changed the designation of November 11 from Armistice Day to Veterans' Day.

That year, after memorial services held by veterans organizations, a massive parade proceeded from Bremerton into the Shipyard, where scores of individuals and families gathered on the pier alongside USS MISSOURI. During an

A fireboat sprays a welcome for USS ST PAUL. Yard tugs, MIMAC, MOANAHONGA, SABEATA and HAST-WIANA join the reception. ST PAUL was the last U.N. ship to leave Korea's Hungnam Harbor on Christmas Eve 1950, following evacuation of troops through that port. The cruiser heads for Pier 5 and three months' overhaul.

Puget Sound Naval Shipyard

Aboard battleship USS MISSOURI, the ship's Commanding Officer, Captain J.R. North, relates details relevant to the plaque on the surrender deck. Interested listeners were Joseph J. King, Industrial Relations Assistant and member of the Veterans Day committee, William O. Wesseler, Master Shop 51, and Committee Chairman Walter Bruns, Master Shop 72.

Kitsap County Historical Society

Roger Paquette (right) Instructor at the Apprentice School, welcomes Albert W. Nelson, Master Marine Machinist, Shop 38, on opening day of the new Apprentice School Building in Sept. 1954. Paquette, a college instructor taught academics at the school from 1947 to 1980. Nelson's interest was based on his association with apprentices assigned to Shop 38. The School motto, "FOR THE MIND — WISDOM: FOR THE HANDS — SKILL", resulted from an all-Shipyard contest. Roger Paquette

inspiring naturalization ceremony on the ship's Surrender Deck,[43] U. S. District Court Judge George H. Boldt administered the Oath of Allegiance to 80 foreign-born men and women.

On September 14, 1954, SALUTE featured the Apprentice Program and praised John Lindberg[44], "father of the present day Apprentice School." Appointed head of the school in 1916, Lindberg quickly recognized the need for professional academic instruction.

The Bremerton School District assigned teachers to help with the academic part of the curriculum. By 1923, an apprentice could spend six hours a week in classes during the work day. Those receiving satisfactory grades for classes taken outside work hours earned an extra five cents an hour. Graduates received high school diplomas. Arvid Grenstad completed his apprenticeship in 1930:

> . . . [early] *in 1930 there was a big layoff, the work force was reduced to less than two thousand, machinists were especially hit hard. Apprentices could not, by law, be let go, consequently since I was the last machinist* [apprentice] *hired in 1926 I was the last to be laid off, the last lay-off occurred only a few days before I finished.*
>
> . . . *I worked for a year before my time came. During that year as a journeyman, I lived pretty high on the hog. It was in the height of the depression years and thirty-five*[46] *dollars or so a week looked very good to my peers, most of whom were unemployed.*

The Yard was able to hire a high caliber of apprentices because of the scarcity of competition from other industries in the local area. During the 1930s, candidates who received a grade of 99 out of 100 on their entrance exams had to wait several years before entering their apprenticeships.

After 28 years in Building 50, the school moved to the floor above the cafeteria in Building 435 in 1948. It remained there until moving up the hill to brick Building 466 north of the officers' quarters in 1954. The latter move was celebrated on September 15 and 16 with an open house and shop and school exhibits and demonstrations.

Special invitations were sent to persons long associated with the apprentice program. Arthur Holden and Emil Olson represented the earliest classes.[47] John C. Lindberg was there. He had been the Director for 25 years; the Bremerton Sun's editorial "The Story of a Man Who Built a School" was a fine tribute to man and school.

When Lindberg retired in 1941, Emil Remmen served as Director for two years. Former instructor Alva Hugenin held the post during the difficult war years when enrollment rose to 1,000. Lester Moyer was director from 1947 until 1955. Succeeding supervisors were Phil Chandler, Lyle Loften, Charles McGuire, Roderick McIntyre and Dan Haas.[48]

Since 1950, Olympic College teachers have handled the academic instruction, and apprentices receive credit toward an associate degree from the College. In 1972 Anna J. Backman became the first woman to graduate from the program. Now, more than a quarter of the apprentices are women.

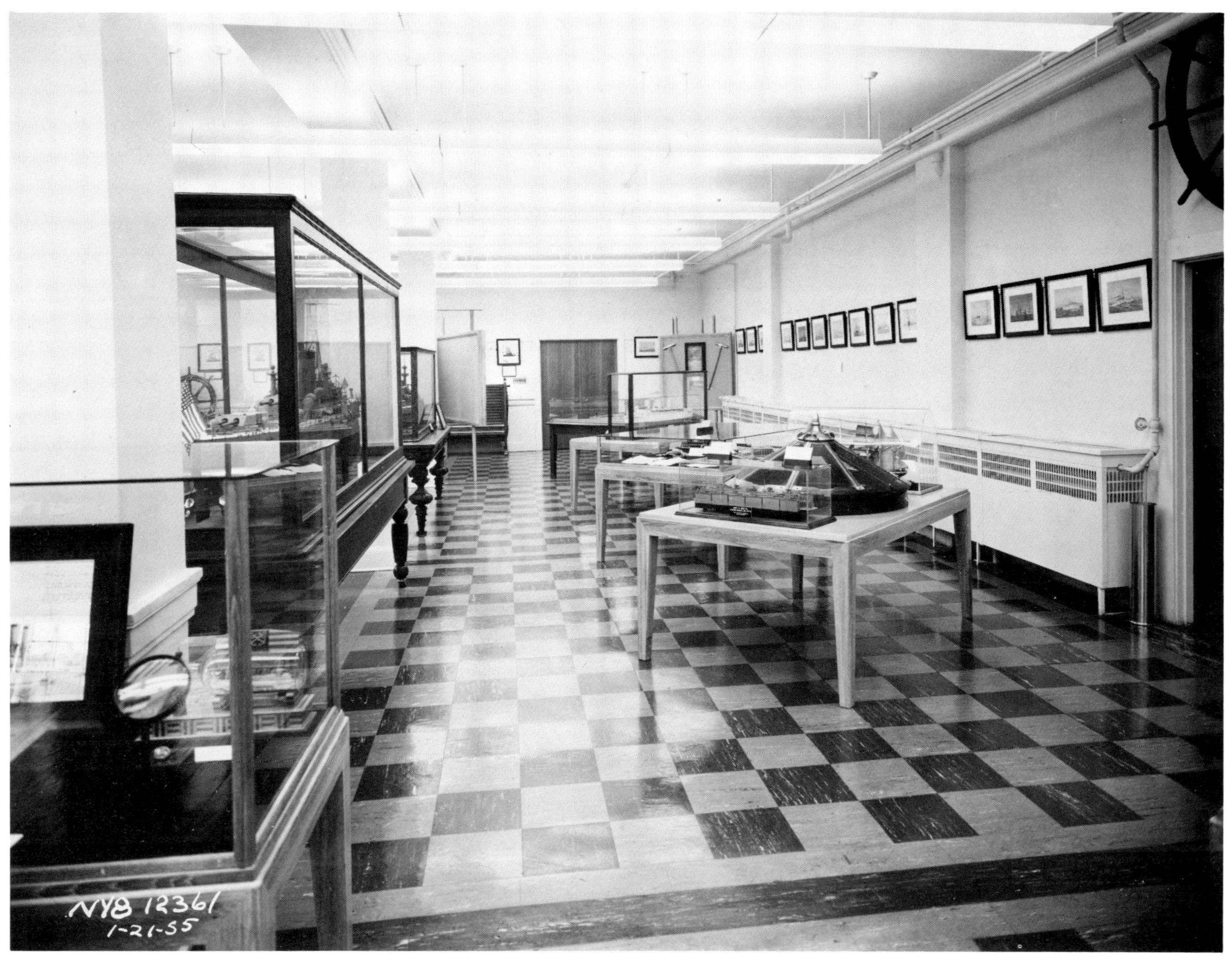

The Naval Museum opened its doors in Craven Center in December 1954. By 1957, Curator Harold Lillehei had catalogued over 1,000 items including rare daguerrotype photos, ship models and guns, cutlasses, ships' figureheads and navigational equipment. That year, over 30,000 visitors from all parts of the nation viewed the exhibits.

Puget Sound Naval Shipyard

When the Shipyard was 60 years old, Shop 07 Plannerman Karl Wood turned in a Beneficial Suggestion regarding the establishment of a shipyard museum. The old five-minute warning bell, which for years had rung from the old foundry in Building 147, lay on a ledge near his work site. His boss, Oliver I. Olsen,[49] Master of the Public Works Shop 07, had a gallery of photographs recording major changes in the Yard. Wood was sure there were many other items suitable for a museum.

The idea was implemented after Rear Admiral Homer N. Wallin[50] became the Shipyard Commander in September 1953. Wallin had established a museum at the Norfolk Navy Yard while he was Shipyard Commander there. With his encouragement, the Puget Sound Naval Shipyard Museum took shape; employees and the public were asked for donations. The SALUTE commented on October 14, 1954:

The past is not dead unless it is left to die. The past is not useless unless we fail to use it. A museum can show us that the Navy is old and mighty — also, that it is a modern institution, still growing, still in change. That we think, is a good thing to show, not only to those who work here but to all those who cross these paths.

In addition to scores of donations from individuals, displays were given or loaned by the Kitsap County Historical Society, the Washington State Historical Society, the Seattle Historical Society and the Museum of History and Industry. Thc Bureau of Ships and the Naval History Department in Washington, D.C., loaned many items.

The Puget Sound Naval Shipyard Museum opened in the northwest corner of Craven Center at Burwell Street and Park Avenue on December 3, 1954. That day, approximately 500 visitors enjoyed the exhibits of items of Shipyard, Navy and maritime interest in the large, well-lit room.

Following the Change of Command ceremony on April 29, 1955, Wallin retired. The new Shipyard Commander was a man well acquainted with the Shipyard and its history — Rear Admiral William A. Dolan.

Carrier CORAL SEA (CVA 43) is under way for sea trials after her major conversion at PSNS from April 15, 1957 to March 15, 1960. This was the largest conversion, measured in man-hours and all other major statistics, accomplished in the Shipyard. Major changes seen in this aerial view include the angle deck, deck edge elevators, new catapults and bridle arresters and a new Pri-Fly, a short term meaning primary flight deck control station. Puget Sound Naval Shipyard

Chapter 7

Carriers, Carriers

1955-1962

In April 1955, a former Navy Yard employee, Rear Admiral William A. Dolan,[1] became Commander of the Puget Sound Naval Shipyard. While a student at Union High School, Dolan worked in the Yard as a rivet passer during the summer of 1919, earning $2.32 a day. The following summer he was a helper electrician third class and earned $3.60 a day. In 1955, he was in command.

That summer, the Bureau of Ships designated PSNS lead yard for the construction of two ships of a new class of guided missile frigates.[2] USS COONTZ (DLG 9) honored Rear Admiral Robert E. Coontz, the Yard's commandant from July 1915 to September 1918 and Chief of Naval Operations from 1919 to 1923. USS KING (DLG 10) was named for Fleet Admiral Ernest J. King, Commander-in-Chief U.S. Fleet during World War II.

These two ships were the Yard's first new construction since the minesweepers AMS 167 and 168 were launched in 1953. On March 1, 1957, retired Master Shipfitter Herman Petersen and Master Sheetmetal Worker V. L. "Vic" Mills[3] struck the symbolic welding arcs which marked the frigates' keel laying ceremony in Dry Dock 3.

Using a new procedure for ship construction, shipfitters prefabricated 30-foot-long hull sections upside down on the shop's assembly slab, permitting down-hand welding. These sections were moved to Dry Dock 3 where they were turned right side up and joined by a continuous welding process[4] to sections already in place. In this manner both hulls grew forward and aft.

In February 1957, PSNS heavy forgers turned out the first of the largest rough shaft forgings ever made on the Pacific Coast. Under Foreman George Cameron's direction, the shop's 2,200-ton hydraulic forging press hammered a 116,000-pound, 46-inch octagon shaped alloy steel billet into a 24-inch diameter forge size with a 40-inch diameter collar. It was 56 feet long.

The rough-forged shaft sections for the two frigates were

The keels for these two new construction frigates were laid in Dry Dock 3 on March 1, 1957. The hulls were built up from 30-foot long prefabricated sections creating the stair-step effect shown. COONTZ (DLG 9) is on the right; KING (DLG 10) is on the left. Puget Sound Naval Shipyard

Nearing completion of the forging process is one of the sections for the DLG 9-10 shafts forged from a high alloy molybdenum-vanadium-carbon steel billet. The steel link chain falls hold the shaft level and are used to rotate it under the 2,200-ton hydraulic hammer. The square block at the end will be hammered into a roughly circular shape and be machined into the shaft coupling flange. The workman in the picture is brushing oxidation slag off the billet.

Puget Sound Naval Shipyard

On a long bed lathe in the Machine Shop, a 31-foot-long intermediate DLG shaft section is undergoing the first rough machining operation. The section of shafting on the right, to the left of the coupling flange, still shows the surface configuration the shaft received from the forge.

Mrs. Fred K. Hicks

An overhead view in the Machine Shop shows the trepanning operation. The trepanning lathe bed is twice as long as the surface finishing lathes on the left. The flat gear drive, seen at the lower end of the trepan bed, advances the cutting head into the rotating shaft at about 45 inches per hour. The work is lubricated with cutting oil supplied through the flexible hose at the lower end of the trepanning tube. Note completed shaft sections in upper right.

Mrs. Fred K. Hicks

moved on railroad cars to Building 538 for removal of heavy scale caused by the forging phase. The shafts then were moved to the Machine Shop for completion by a new process known as trepanning, to remove a solid core from the center of each shaft. This resulted in a major energy savings and as a side benefit produced a core cylinder instead of chunks and shavings. The process attracted nationwide attention for Shop 31. A few private manufacturers used the method on small scale jobs, but trepanning of a propulsion shaft had never been done before in a U. S. naval shipyard.

In a post-completion technical report describing the work, Shop Master Fred K. Hicks[5] wrote:

The trepanning process of deep hole drilling represents the most significant "breakthrough" in machining since the advent of carbide cutting tools. I can think of no other single major item of ships' components which has undergone such vast improvement and acceleration in manufacturing methods.

In preparation for coring, the shaft was placed in dead axial alignment with the axis of the trepanning tube by adjustment of the steady bearing roller assemblies. The trepanning cutter head cut an annular ring 11.750 inches in diameter through shafts over 50 feet in length, leaving the core intact. This resulted in a great savings in time; four shafts could be cored in the time formerly taken to bore one. Also, the core cylinder could be sold at a higher price than chips resulting from the spade method.

SALUTE described the operation on February 21, 1958:

. . . The shaft section is mounted so one end is open to the special tool mounted on the lathe extension bed . . . The tool resembles a polished steel pipe about a foot in diameter and nearly 80 feet long. Its business end is equipped with a special tungsten carbide bit a little wider than the thickness of the walls of the pipe like tool. As the "carboloy" bit is fed into the rotating shaft section it shaves away small pieces of the shaft metal.

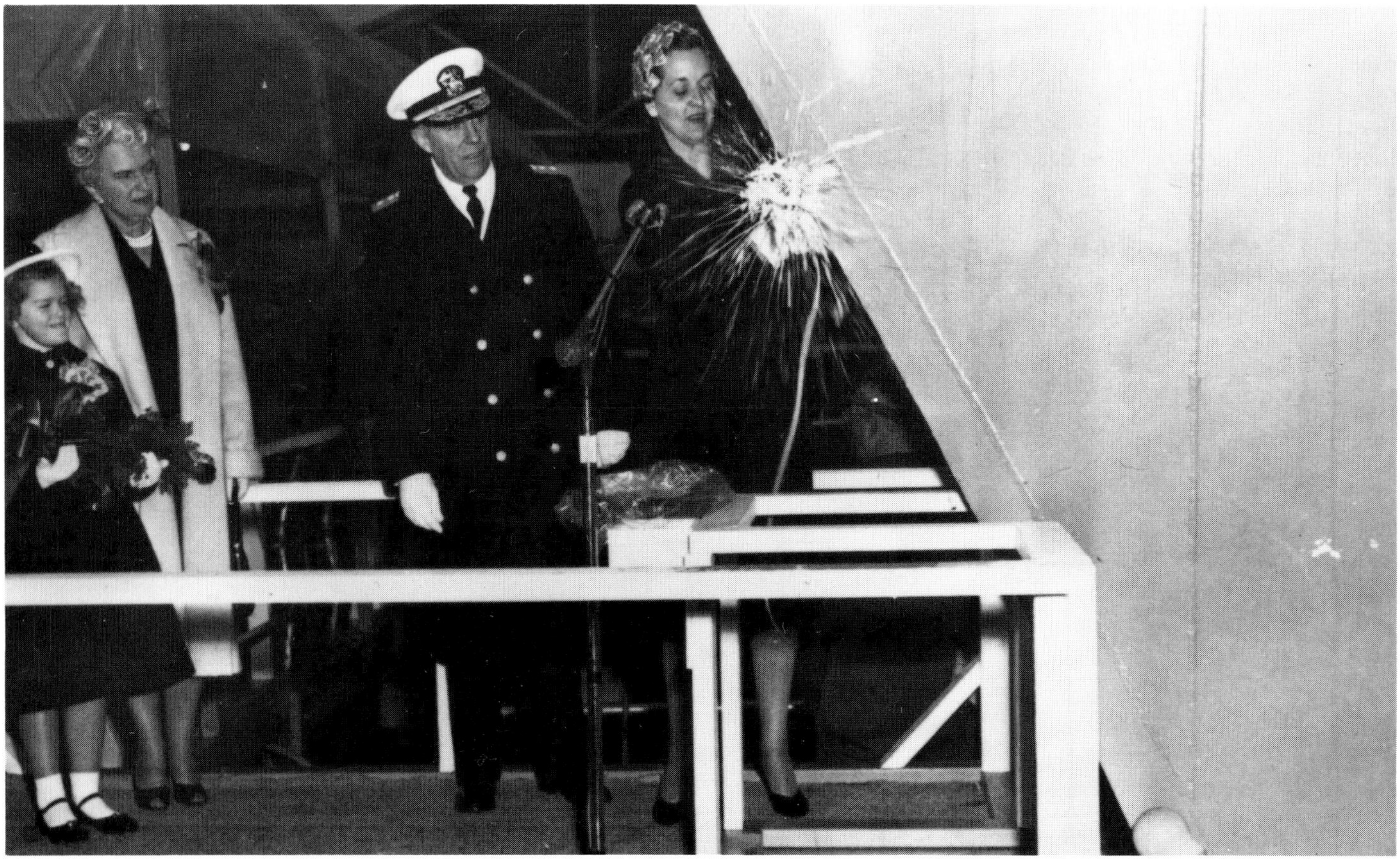

USS COONTZ (DLG 9) and USS KING (DLG 10) were christened in a dual ceremony in Dry Dock 3 on December 7, 1958. In the upper photo Mrs. Robert J. Coontz, wife of the grandson of the late Admiral Coontz, christens COONTZ while Rear Admiral Frank T. Watkins looks on. Also watching are Matron of Honor Mrs. Edwin Kokko, who as Bertha Coontz lived in Quarters C when her father was Commandant, and Flower Girl Barbara Diane Schreiber. In the lower photo Rear Admiral Phillip W. Snyder watches Mrs. Oliver van den Berg, daughter of the late Admiral Ernest J. King, christen USS KING. Not shown are Mrs. James D. McReynolds, Matron of Honor, Elizabeth van den Berg, Maid of Honor and Ellen van den Berg, Flower Girl. *Puget Sound Naval Shipyard*

The Boiler Shop under Master Richard P. "Rip" Pendras built the eight boilers for COONTZ and KING. With operating pressure of 1,200 pounds per square inch, these boilers were the highest pressure boilers built by the shop to that date. In May 1958, the first two of the approximately 1,700-tube boilers were delivered to COONTZ.

As a steady stream of prefabricated sections moved from the Shipfitters Shop to Dry Dock 3, two new ships took shape. Shipfitters, welders, pipefitters, riggers, electricians and marine machinists swarmed over the ships.

On a wet December 6, 1958, the two frigates were christened. Vice Admiral Ruthven E. Libby, Commander First Fleet, praised the "can-do" spirit of the people of the Yard that made possible PSNS's position as "keystone in the defense of the country."

New 50-star[6] U. S. flags flew from the fantails of the frigates at their commissionings. Commander Herman H. Ries took command of USS COONTZ on July 15, 1960, and Commander Melvin E. Bustard of USS KING on November 17, 1960.

This new class of ship was fast and powerful, incorporating all the latest developments in communication facilities, improved air conditioning systems and improved habitability for the crew. The ships, which had the latest in missile and fire control systems, were armed with the new surface-to-air Terrier missile and equipped with the first anti-submarine rockets, as well as the newest sonar system.

Speakers at the launchings and commissionings of the two ships praised this weaponry, especially the surface-to-air guided missile. Powered by a solid fuel rocket motor, the Terriers made the frigates formidable warships. They were designed to intercept enemy aircraft at longer ranges and higher altitudes than conventional antiaircraft guns.

Work on specialty ordnance systems was not new at PSNS. The Ordnance Design section began forming in the 1930s under Henry Hitt. By 1939 the group was tasked with its first special project, the mounting of an optical rangefinder and an antiaircraft gun director on a single mount for tracking aircraft and controlling 1.1-inch machine guns. The director was revised from 5-inch gun characteristics and the assembly was installed on USS ARIZONA.

Mark 8 Missile Launcher is being assembled in Building 500 as part of the Guided Missile Launching System (GMLS) Mark 11. The entire assembly, with 15 missiles in the circular magazine below the launcher and a missile on each launcher arm, collectively weighing 230,000 pounds, was subjected to acceptance testing on this first stage before being shipped for shipboard installations.

Puget Sound Naval Shipyard

The team continued to grow, attracting high caliber design engineers, who did all types of ordnance system design; electrical, mechanical, optical and structural, as well as the extensive documentation required. Recognition of the group's expertise and dedication came in 1946 with the presentation of the Navy Development Award for Outstanding Research and Development.[7]

The group designed and successfully tested a prototype automatic loading system for rapid fire of one of the three eight-inch guns in a CA class cruiser turret. However, as Richard Linkletter explains:

> *A projectile fired from a gun has no ability to change course once fired, . . . A missile properly equipped can follow the changes of the target's path like a hunting dog and close in for the hit. After the kamikaze experience of World War II, the Navy was all for hunting-type projectiles.*

The unique Ordnance Design group at PSNS was tasked to design, build and supervise installation of Missile Launcher X-5 on the experimental ship, USS NORTON SOUND (AVM1). For this the Shipyard received from the Bureau of Ordnance a Letter of Commendation for the first remotely controlled guided missile launcher in the free world.[8]

Other projects followed in rapid succession. Linkletter remembers the twin-arm launchers with improved elevation drives the group designed, installed and tested on USS MISSISSIPPI:

> *This project was completed for the miniscule cost of $320,000, a figure used by the Bureau of Ordnance for years to show contractors how inexpensively such systems could be built.*

While still involved in three major projects in 1955, Ordnance received the order to design and build a prototype and 19 units of the Guided Missile Launching System Mark 11 for the Tartar Missile.

To handle the expansion of work load, the roster of Ordnance Design increased to about 70 Civil Service and 70 contract engineers. Eugene Hoffman was the civilian in charge as Assistant Chief Engineer, Ordnance.

Tartar was intended primarily for installation aboard destroyers, but was also installed aboard cruisers as secondary missile batteries. Tartar missiles launchers were also installed on IWO JIMA (LPH 2), which was built in Dry Dock 3 after COONTZ and KING had been launched.

IWO JIMA was a new class of ship. BuShips designated PSNS lead design yard for this amphibious assault ship and also assigned construction of IWO JIMA (LPH 2)[9] to the Shipyard. This marked the beginning of a new era of ship design.

ORDNANCE DESIGN, PUGET SOUND NAVAL SHIPYARD FEB. 4, 1946
PRESENTED: NAVAL ORDNANCE DEVELOPMENT AWARD FOR OUTSTANDING RESEARCH & DEVELOPMENT
BUTTERFIELD BOSSLER WINSOR OLSON HITT, P6 CAPT V.B. COLE DESIGN SUPT SCHAIRER SHERMAN BISHOP, P5 MERETH P4 EITEL
ROUCH MARTINSON CROSBY HERRELL WOODIER CH D. HULL WAGNER SELMENSKY, P4 HAGBO, P4
EATOUGH SULLINS GOORE P4 DAVIS SCHAFER SECY ASH HAINES HOFFMAN, P5 HLAVIN
ABSENT: STROSHINE, ANDERSON DETACHED COLBY, JOHNSTONE LAUDERBACK

Kenneth Johnson[10] became Chief Design Engineer in 1954 and began a reorganization of the Design Division to meet increased demands. Work was divided into groups, according to engineering discipline, to spread the heavy work load. The work load co-ordinating group was greatly expanded to maintain better cost control.

To meet the increased work load, members of the Design Division began visiting all high schools in the Puget Sound area to recruit graduates as Draftsman Trainees, GS-2, at a starting salary of $2,450 a year. Engineering students from colleges throughout the Pacific Northwest received on-the-job training.

In preparing the contract design for IWO JIMA, the Yard identified and obtained Bureau of Ships' approval for 82 advance changes to the preliminary design. This resulted in almost $3,000,000 savings in the construction of the three ships of that class.

By August, Master Molder John E. Langhans reported the Foundry's completion of the first of seven sections for IWO JIMA's 90-ton stern frame. The Foundry also cast the seven-section stern frame for LPH 3, which was being built at the Philadelphia Naval Shipyard.

Work on this, the first Navy ship ever built from the keel up for helicopter use, continued for two years. IWO JIMA was christened September 17, 1960, with Mrs. Harry Schmidt as sponsor. Captain Thomas D. Harris became commanding officer, when the ship was commissioned August 26, 1961.

IWO JIMA (LPH 2), the first of a new class of amphibious ships which was designed and built at PSNS, takes shape in Dry Dock 3. The almost completed flight deck rises above the staging. The ship's bulbous bow looks out through the staging forest on the drydock floor. IWO JIMA was launched and christened on September 17, 1960.

Puget Sound Naval Shipyard

Heavy cruiser COLUMBUS (CA 136) was converted to guided missile cruiser COLUMBUS (CG 12) by removing her 8-inch main battery and all of her superstructure, and replacing it with Talos missile launcher systems, fore and aft, Tartar missiles, port and starboard, and an aluminum bridge structure with sophisticated electronics equipment. To save top-side space and weight, the old masts and stacks were combined into primarily aluminum Macks. Puget Sound Naval Shipyard.

Test firing of two Talos missiles from a shipboard deck-mounted twin arm launcher. Visible through the smoke are the two saucer shaped target illumination and tracking antennas and the antennas for the missile guidance radars.

Larry Jacobson

From March 1959 until early in 1963, the Yard was involved in the conversion of USS COLUMBUS from heavy cruiser (CA 136) to guided-missile cruiser (CG 12). In operation "Clean Sweep", shipyard workers removed COLUMBUS' guns, turrets, and the entire superstructure down to the main deck, leaving only the hull and the engineering plant.

This conversion also offered the opportunity to improve crew's quarters aboard ship. On November 30, 1962, SALUTE reported:

> *Enlisted berthing spaces are greatly modernized for habitability. Bunks are no longer stacked in tiers of four and five. The majority are in double* [two high] *bunks with a few in tiers of three. Lockers are not to be seen in the compartments. The tops of the bunks lift up to reveal each man's locker beneath his foam rubber mattress.*
>
> *Bunks are separated by aluminum partitions to afford more individual privacy. Each bunk also has an individual reading lamp. All berthing compartments have adjacent toilet and shower facilities. And, perhaps best of all, from the crew's standpoint, all working and living spaces are air-conditioned.*

At her recommissioning on December 1, 1962, the ship's new armament consisted of Talos and Tartar Missiles and ASROC.[11] Her deck bristled with missile launchers, radar antennas and "Macks", the latter a space-saving combination of masts and smokestacks.

Although PSNS worked on many types of ships during the 1950s, the decade is probably best remembered for aircraft carrier conversions. Editor Jim Reems said it well in SALUTE on January 22, 1960:

> *. . . an era of carrier conversion that began modestly back in the winter of 1948 with the pulling out of the mothball fleet of ESSEX, 27,100-ton carrier of WW II vintage, for limited modernization, and which ends 12 years later with the return to the fleet of CORAL SEA, a massive 65,000-ton postwar creation into whose hull they* [the Shipyard workers] *have put the latest in carrier refinements that will give her operational capabilities second to none.*

An unprecedented number of carriers were assembled at PSNS in August of 1955. Shown here are LEXINGTON at Pier 6, SHANGRI LA and MIDWAY at Pier 5, ESSEX in Dry Dock 4, YORKTOWN at Pier 3 and FRANKLIN D. ROOSEVELT in Dry Dock 5.
Puget Sound Naval Shipyard

After her undocking, SHANGRI LA's conversion was increased in scope to include a hurricane bow, a modernized Pri-Fly and the newest arresting gear. The work was done at Pier 3 where she is shown facing seaward for later dead-load catapult tests in 1954.

Puget Sound Naval Shipyard

Between ESSEX and CORAL SEA, there were KEARSARGE, YORKTOWN, HANCOCK, LEXINGTON, SHANGRI LA, FRANKLIN D. ROOSEVELT and MIDWAY, nine in all, and each succeeding conversion, in turn, received added physical improvements that made it even more modern than its predecessor.

In the beginning, aboard ESSEX, KEARSARGE and YORKTOWN, there was the addition of blisters for added protection and increased stability, the broadening of flight decks for handling of heavier, faster planes of the new jet age, increased fuel facilities and special facilities for handling and stowage of strange new weapons, jet blast deflectors and oxygen-nitrogen systems, and improvements in crew berthing spaces . . .

Halvor Halvorsen,[12] head of the Design Structural Division at that time, later described the work thus:

We were always thinking "big" in order to develop new concepts and reduce construction costs. One of the earliest items that comes to mind is the installation of saddle gasoline tanks during the CV 9 class conversions . . . In lieu of removing the original gasoline tank structures and constructing new tanks piece by piece within the ship at the expense of an inordinately long dry dock period . . . The new tank structures were prefabricated in the Shipfitters Shop prior to ship's arrival with master butts [joints] *strategically located. The original tank structures were removed thru a large opening in the ship's bottom. The new tanks, some sections weighing over 20 tons, were lifted in place thru the holes. Upon completion of welding the ship's bottom was reinstalled as a plug and the structure was complete.*[13]

Reems' article continued:

The Shipyard's hammerhead crane dominates the skyline with carriers SHANGRI LA (CVA 38) and YORKTOWN (CV 10) left and right at piers. The bow of battleship ALABAMA (BB 60) is seen center right. SHANGRI LA shows her new angle deck. The angle deck change was not included in YORKTOWN's first conversion. Larry Jacobson

Confronted with an onrushing jet age in which development of heavier and faster planes was so rapid that it already threatened the Navy with early obsolescence in carrier modernization even before completed, with HANCOCK, Navy sprung its first in spectacular modern carrier refinements, steam catapults.

First steam catapults ever to be installed aboard an American warship, for PSNS, now designated as Lead Design Yard in 27C conversions, the installation of this new phase of modernization aboard HANCOCK made this Naval Shipyard a center of [national] *attention.*

As planes became heavier, the design and motive power of catapults changed. SARATOGA and LEXINGTON were built with flywheel catapults. Compressed air, black powder and water were also used as propellants; none could be upgraded to handle the heavy jet aircraft.

The first shipboard steam catapult was installed late in 1948 aboard HMS PERSEUS. The United States purchased design rights from Great Britain and began studying the problems of increasing catapult power and capacity.

L. L. "Larry" Finnegan,[14] head of the Design Division's Marine and Mechanical Subdivision, traveled to England to discuss the installation of steam catapults on carriers with England's top civilian and naval designers.

Upon his return, Finnegan organized the steam catapult design crew: Branch heads Carl L. Newstrom of Scientific and Test, Carl Robbie of Marine and Mechanical Engineering, Alexander Torbitt Jr. of Piping and Ventilation, and Halvor Halvorsen of Structural and Fitting. Principal engineers were Earl Lagergren, Neil Lardy, Jack Bollen, J. H. Fyten, Jack Alguard and Willard Lynch.

The first operational test of a steam catapult on an American ship occurred February 8, 1954. Under a dull gray sky, a massive "dead load" was fired from USS HANCOCK (CV 19) into Sinclair Inlet. The "steam slingshot" was a success.

Generating six times as much power as hydraulic cata-

KEARSARGE is being fitted with a new deck-edge elevator similar to those installed in other conversions. Careful coordination among the riggers, crane operators and marine machinists was required for the successful placement of the elevator.
Puget Sound Naval Shipyard

pults, the new device made it possible for carriers to launch heavier planes under more adverse conditions of wind and sea. Yard personnel were aboard HANCOCK during her six-month shakedown cruise, fine tuning the new system so it operated as successfully with half-a-million dollar jets as it had with rubber-tired test loads.

Reems continued in the SALUTE article:

For PSNS and its shipworkers, this conversion reaped recognition in having made a contribution to Navy carrier launching history; with HANCOCK's conversion also came top Shipyard management's word that "we are now participating in one of the most extensive conversions of capital ships ever undertaken by the Navy . . ."

By now, four carrier conversions in less than six years developed a corps of conversion designers and artisans to whom the new developments in carrier modernization were welcome challenges that would further test and increase their savvy and know-how. This veteran corps of conversion specialists was to be given the full treatment during the modernization of SHANGRI LA . . . when a number of structural bombshells [sic] *fell . . .*[15]

Already heading home in completion, SHANGRI LA was in the throes of final compartment closures when the first bomb fell. Bureau of Ships notified the Yard that CVA 38 was slated for the newest in carrier silhouettes, that is, she was to get the angle deck. SHANGRI LA became the first carrier to receive a permanent version of the angle deck which had been given a trial installation aboard ANTIETAM.

An angle deck is an extension of the ship's flight deck, jutting off to the portside opposite the island. This provides a landing path separate from that used during launching operations.

Of particular interest in SHANGRI LA's conversion was the ship's deck-edge elevator. Installed on the ship's port side, the elevator, when in a raised position became an extension of the new angle deck. Constructed entirely of aluminum, the 87,000-pound framework weighed 34,000

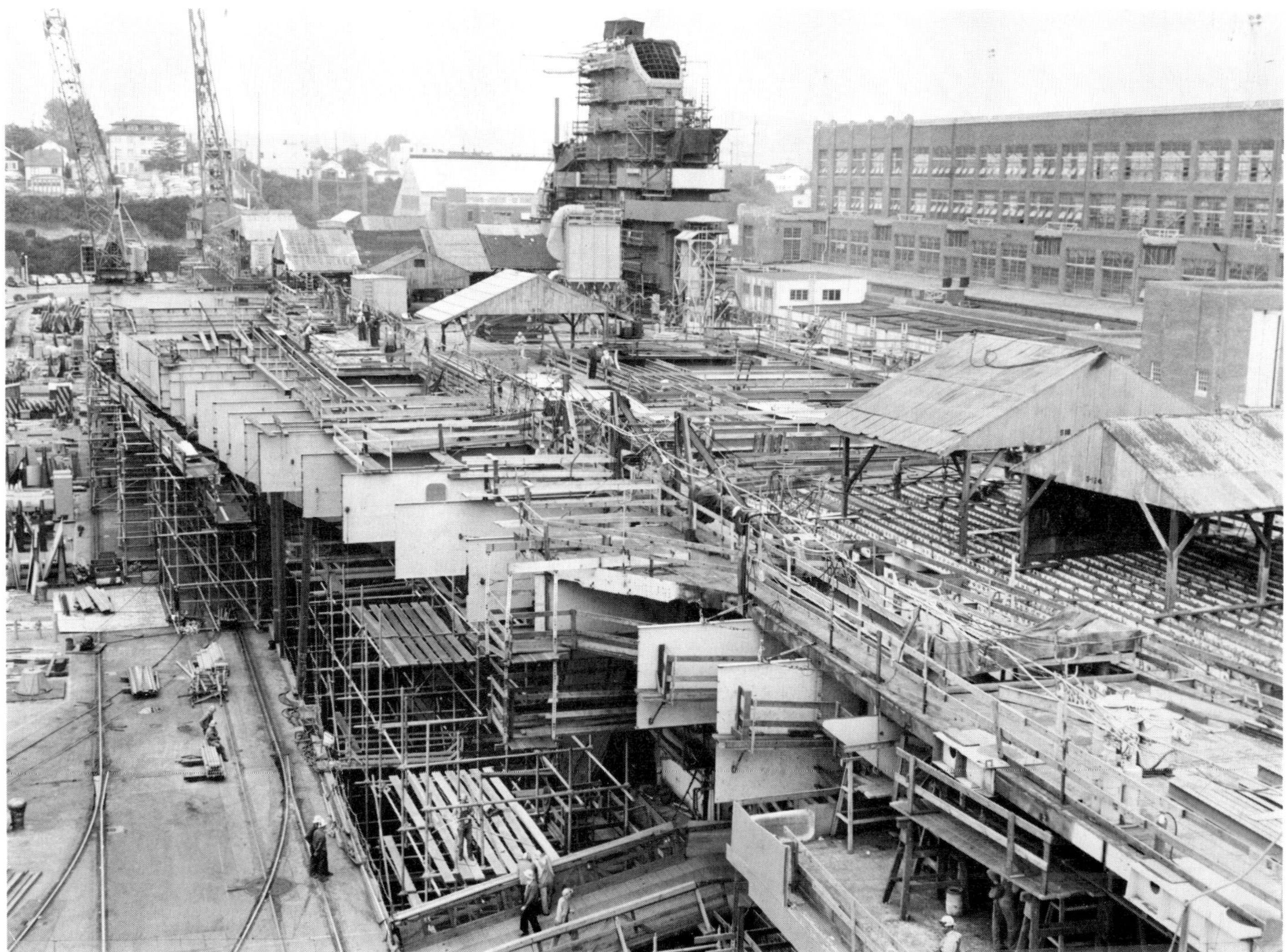

LEXINGTON (CV 16) in Dry Dock 4 for the major structural work associated with installation of the angle deck and the deck-edge elevators. The structural framing extending out from the flight deck will carry the weight of the new deck. The portable equipment inboard of the island is a dust collecting unit to collect spent grit from sandblasting.

Puget Sound Naval Shipyard

pounds less than the steel elevator it replaced. The load capacity increased from 46,000 to 57,000 pounds.

Reems continued in the January 1960 article with word on the other two bombshells:

Shipyard conversionists had by now moved SHANGRI LA out into open water, complete with conventional bow, when BuShips asked for a hurricane bow, streamlined and affording greater protection against the elements. Shipworkers tackled fabrication and installation of the hurricane bow and were not surprised when blueprints appeared calling for a greatly modernized primary flight control booth. Replete with her new angle deck, steam catapults, hurricane bow, modern Pri-Fly,[16] the world's largest all aluminum deck-edge elevator, plus the latest in arresting gear and barricade changes, improved fuel and provision facilities, SHANGRI LA's recommissioning day was January 10, 1955.

Five down and four to go and 1955 was to be a big year with LEXINGTON almost completed, FDR well under way, and MIDWAY headed for PSNS around the Horn and due to arrive in early August. Already strung along the waterfront were FDR, LEXINGTON, SHANGRI LA, ESSEX and YORKTOWN, the latter two in for seven-month modernizations, known as "baby conversions", to receive the hurricane bow, angle deck, and improved arresting gear . . .

At the beginning of her availability in May 1954, USS FRANKLIN D. ROOSEVELT's conversion was speeded by the removal of her armor plate while the ship was still waterborne. Master Rigger Walt Bruns devised the plan of lowering the armor plates onto a heavy barge secured to the ship's hull.

Catapult testing using a massive "dead-load" vehicle to simulate an aircraft launching was a critical operation in certifying the performance of a new steam catapult. Pressures, catapult shuttle velocities and the trajectory of the dead-load were measured to calculate the catapult's launching power. The upper photo shows a dead-load being propelled off BENNINGTON's flight deck. The lower photo shows the splash after a launch from BON HOMME RICHARD.

Grosso Collection/Kitsap County Historical Society

Black smoke rises from the stack of LEXINGTON (CV 16) as she leaves the Yard with her ship's company at quarters in September 1955 after her conversion. Other ships are USS MIDWAY (CVA 41) at Pier 5, and USS FRANKLIN D. ROOSEVELT (CVA 42) at Pier 3.

Puget Sound Naval Shipyard

Senator Warren G. Magnuson inserted into the Congressional Record on June 28, 1954, a quote from SALUTE's June 11 issue. The paper quoted ship superintendent Lieutenant Commander E. J. Bloom as follows:

I've seen nothing like the present spirit shown to date on the FDR. To me, the individual worker's morale is at the highest peak I have known on a conversion job. Master Walt Bruns' riggers seemed to catch fire almost from the moment they started on the plate removals and believe me, the Shop 11-26 multiple force of shipfitters, chippers and calkers, burners and drillers were crowding them all the way.

Puget Sound Naval Shipyard was assigned the conversion of all three MIDWAY class carriers. MIDWAY (CVA 41) arrived in August 1955 bringing to six the number of carriers at PSNS.

Reems' article continued:

PSNS continued to incorporate into each succeeding conversion the latest in modern carrier design, to the end that when FDR was recommissioned in 1956 and MIDWAY in 1957, the feeling of worker majority was that considerable progress had been made.

Installation of steam catapults, angle decks and hurricane bows along with greatly improved Pri-Flys, CIC areas and crew berthing spaces, had become standard conversion procedure. Such additions as deck-edge elevators, escalators, tougher and stronger arresting gear and barricades, wet accumulators and increasing of maximum beam amidship now were considered to put FDR and MIDWAY in a similar class to FORRESTAL — maybe not as big, but every bit as operationally efficient.

The major part of the conversion on CORAL SEA was done in Dry Dock 4. Work on the installation of the steam catapults was protected from the weather by the two rows of sheds on the bow. The structural work on the angle deck, which overhangs the west side of the dock, is nearly complete. Grosso Collection/Kitsap County Historical Society

An A3D Skywarrior tested MIDWAY's new water-cooled jet-blast deflectors in April 1958. These were designed at PSNS to deflect the jet's exhaust blast upward to allow greater safety for personnel and aircraft in the vicinity of the catapults.

In a later issue of SALUTE,[17] Reems continued:

March [1960] *saw the completion of the largest conversion job ever undertaken by the Yard. Reference is made, of course, to the CORAL SEA, a $75,000,000 effort. Major changes involved in the conversion were the broadening of the vessel's beam, installation of new deck-edge elevators and a new angle deck, and modifying ammunition magazines to accommodate modern weapons. Conversion was begun in April 1957.*

A feather in the Shipyard's cap was the commendation received from the Bureau of Ships for turning back in excess of $1,500,000 on this project.

One of the most eagerly awaited ships in the 1950s was USS BREMERTON (CA 130). On Thursday, July 25, 1957, BREMERTON arrived at the Yard for a four-month regular overhaul availability. The Thirteenth Naval District Band played and the crowd cheered as the ship moored alongside Pier 5. That Saturday and Sunday the public was invited to visit the ship. On Saturday night Pacific Avenue was roped off between Fourth and Fifth Streets for a dance to honor the ship's crew.

BREMERTON also spent a day in the Yard in August 1959 while she was in the area for SeaFair festivities in Seattle. On both occasions, the ship's sponsor, Betty McGowan,

Under way after her conversion, FRANKLIN D. ROOSEVELT shows off her major new features: bow catapults, jet blast deflectors, bridle arresters, angle deck and modernized Pri-Fly. Puget Sound Naval Shipyard

PSNS's giant hammerhead crane makes its second lift of a 104-ton transformer, June 1955, from the freighter VANCOUVER STAR. At that time no other local crane could handle the load. This transformer was the second of three to be landed here and then sent by rail to the new Duwamish substation in Seattle. Puget Sound Naval Shipyard

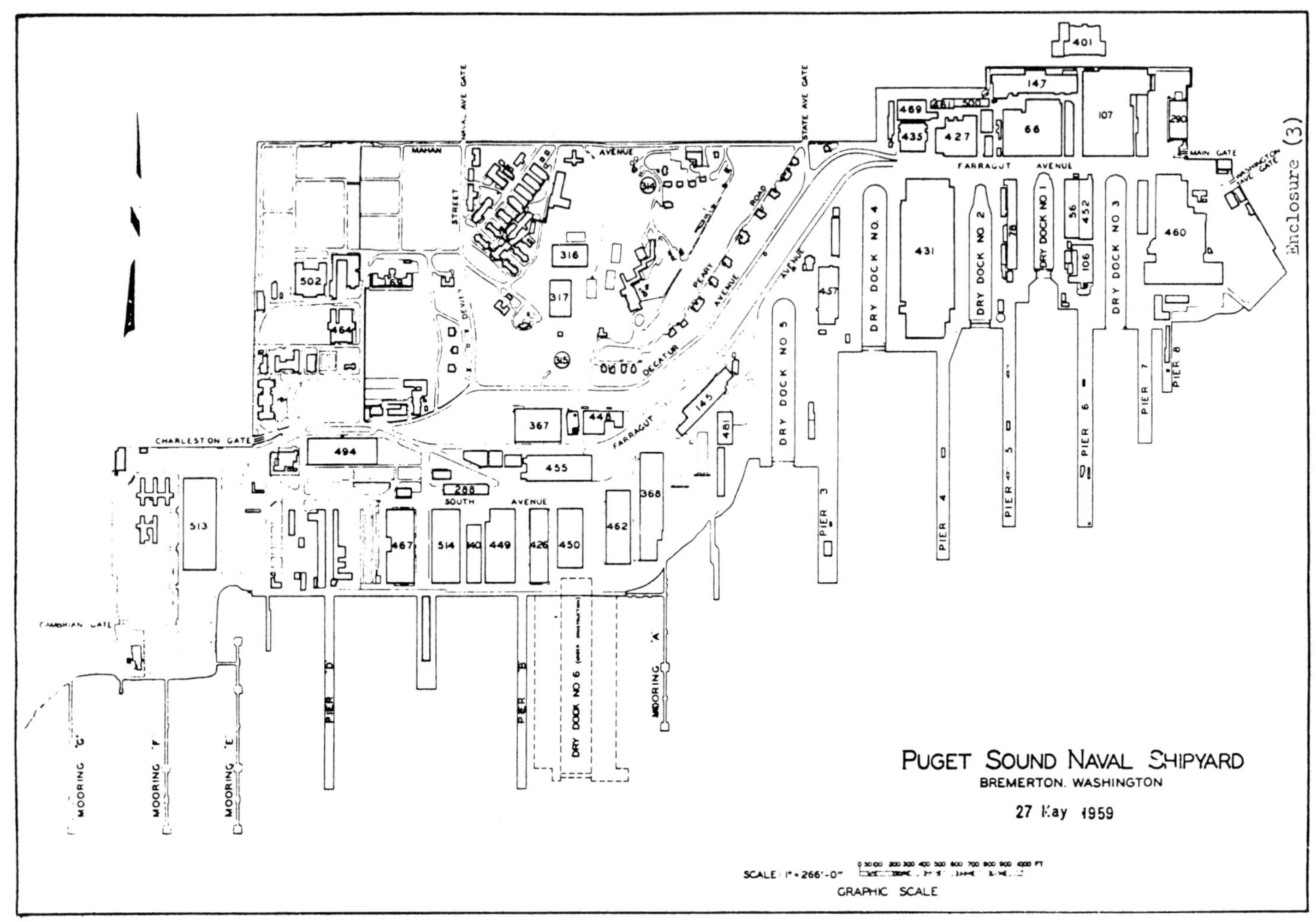

In 1958, Puget Sound Naval Shipyard contained 316 acres of land with 3 miles of perimeter fence, 16 miles of roads, 24 miles of crane and railroad trackage, 249 buildings, 13 piers and 5 dry docks. The following year, work began on Dry Dock 6, as indicated in the lower mid-section of this map. On this map, but not the one on page 116, are the Reserve Fleet moorings in the west end of the Yard.

Puget Sound Naval Shipyard

was part of the official welcoming committee.

In special cases, with Navy Department approval, Naval Shipyards have made their machinery and workers' abilities available to private industry and other government installations for critical jobs when no commercial facilities were available. In 1955, PSNS's hammerhead crane unloaded from the freighter VANCOUVER STAR, transformers for the new Duwamish Substation in Seattle. Shop 31 machined a giant gear blank for a Portland firm in 1957 and also produced a 24-foot diameter mold ring for a California aluminum plant. Inside Machinists, under Leadingman Mel Wortman,[18] removed, machined and replaced the wearing rings of a large hydraulic turbine runner for Seattle Light's Ross Dam.

The July 1958 issue of the BuShips Journal featured Puget Sound Naval Shipyard. It gave the history of the Shipyard, stressing PSNS's pioneering developments in many fields. It listed "recent examples": cable banding for ship electrical cable installations, optical tooling for checking machinery accuracy, inexpensive filler (popcorn) for plastic patterns, trepanning of DLG shafts, and pouring special alloy AL 220 castings.[19]

PSNS's interest in new and modern techniques was not limited to industrial use. The Journal article termed PSNS's administrative achievments equally impressive. PSNS was the first naval shipyard to utilize electronic data processing equipment, having installed an IBM 650 computer in 1956. The article noted the computer was used "daily" in solving complex engineering problems.[20]

The Comptroller Department spearheaded the development of specifications for the installation of a computer in their department, but it was also used in the Production Department.

Production Analyst Roy Workman wrote:

> *The Production Officer told me to . . . attend an IBM programming course in Seattle and convert our production scheduling and control system . . . to the IBM 650. When we received our IBM 650, we had an embryo system ready to go.*

Walter Bruns, Master of the Riggers and Laborers Shop, reviews the apprentice training program in June 1960 while Fred Timmerman, Bruns' successor as Shop Master looks on. Bruns is seated in the chair used by his predecessor, George Trahey. The chair shows the repair of a hole where a .32 caliber bullet passed through to pierce Trahey's shoulder when, in 1912, a crazed employee tried to kill him.

Puget Sound Naval Shipyard

The first PSNS employee awarded the Distinguished Civilian Service Award[21] was Foreman Machinist Carl Forsmark[22] for his mastery of technical machinery problems. The second was William O. Wesseler[23] in 1955 for his work on cable-banding and shop planning. Captain David R. Saveker, who was Acting Repair and New Construction Superintendent in the late 1950s, described Wesseler thus:

He was an innovative, hard charging and decisive leader in the electrical/electronics area. He had a keen analytical bent and saw that X51 kept accurate and detailed production records of the shop's activities. As a result PSNS led the naval shipyards in electrical work planning, and initiated electrical planning standards that set the pace for ship construction, conversion and repair.

Walt Bruns[24] received the award in 1956 for advancing the state of the art of rigging. At the outbreak of World War II, Bruns, then a Foreman Rigger, began a program to train helpers and laborers in proper rigging procedures, previously jealously guarded, in order to increase the number of qualified journeymen riggers.

He was appointed Master of Shop 72 in 1944. Protective of the rights and well-being of his work force, Bruns succeeded in establishing the rating of apprentice rigger. By proving the skill necessary to handle complicated and dangerous work assssignments, he was instrumental in raising riggers' pay to the level of the "bench mark" trades.[25]

Fred Timmerman, who became Master of Shop 72 when Bruns retired in July 1960, considered Bruns:

. . . one of a kind, one of the last of the old time master mechanics who exercised almost total control of their domain. They were the major source of trade knowledge and know-how — they were the backbone of the Naval Shipyards.

Many men, who started work in the Puget Sound Naval Station/Navy Yard during its earliest years, retired with 40 or more years' service. However, the increased employment during World War I was reflected in the larger number of

Breaking ground for the construction of Dry Dock 6, Rear Admiral Frank T. Watkins drops a wrecking ball to begin crushing the pavement at the dock site. Photographer Stan Cleary records the start of this three-year project.

Grosso Collection/Kitsap County Historical Society

In 1959, the Building Ways at the west end of the yard next to Buiilding 480, were razed in connection with the construction of Dry Dock 6. The Woodworking Shop, (Bldg. 851) is now located in this area.

Grosso Collection/Kitsap County Historical Society

Rear Admiral Eugene J. Peltier, Chief of the Bureau of Yards and Docks, and a team of engineers look over the excavation for Dry Dock 6. Vehicular traffic in and out of the pit traveled over the ramp at the left of the photo.

Grosso Collection/Kitsap County Historical Society

Pouring concrete for the floor of Dry Dock 6 required careful placement of the concrete hopper for each pour. The size and strength of the dock's structure is dramatized by the size of the supporting beams and by the diameter of the installed reinforcing bars.

Grosso Collection/Kitsap County Historical Society

To commemorate the completion of Dry Dock 6 and to signal its readiness for operations, Senator Henry M. Jackson unveils the memorial dedication plaque. Shipyard Commander William A. Dolan stands at his side.

Grosso Collection/Kitsap County Historical Society

''senior'' employees during the 1950s and 1960s. Many proudly wore pins indicating 40 years' service.[26]

The first woman to receive a 40-year pin was Genevieve Wolfe.[27] In 1949 she had become the first female GS-13 in the Thirteenth Naval District. When she retired in 1957, the former Yeomanette was head of the Comptroller's Accounting and Disbursing Division. That November, Helen V. Miller[28] also retired, the last of the World War I enlisted women working at PSNS.

A major organizational change took place during 1960. In the Production Department, all shops were divided into five groups according to the type of work they performed, with a Group Master at the head of each group. The new Group Masters were: Dean C. Calhoun, Structural; Edward S. Moskeland, Machinery; Eugene H. Tennyson, Outfitting; William O. Wesseler, Electrical; and Fred L. Timmerman, Service. The following year Lloyd Welch became Master of the Public Works group.

Later, Outfitting was integrated into the Structural and Machinery groups. The designation ''Master'' was changed to ''Superintendent''.

The Production Department itself was reorganized to conform with directives from the Bureau of Ships, into the Repair Division, the Shipbuilding Division and the Production Engineering Division.

On May 27, 1960, Dolan again became Commander of the Puget Sound Naval Shipyard. In an unusual command cycle, he relieved Rear Admiral Phillip W. Snyder, who had relieved him in 1958. That July, U.S. Naval Base, Bremerton, was disestablished. The Naval Barracks came under Shipyard command; the other satellite commands reported directly to the Commandant of the Thirteenth Naval District.

On December 7, 1961, Admiral Dolan dedicated a new steel blasting and painting plant in the west end of the Yard. The new plant could process up to 40 feet of plate a minute, handling plates up to 10 feet in width, 40 feet in length and 2 inches thick. This was in keeping with the Yard's policy of adoption of new methods of handling steel to meet the demands of the Yard's heavy workload.

The major addition to the Yard was a new dry dock. On Christmas Eve 1958, the Navy announced a well-deserved addition to PSNS, a dry dock large enough to hold the new Forrestal class carriers.

Tugs assist KEARSARGE (CVS 33) into Dry Dock 6 on April 23, 1962 for the dock's dedication and first docking. To honor the occasion, ship's company, on the flight deck, line the rails and spell out "PSNS DD-6". Puget Sound Naval Shipyard

The following year Building 480 became the nerve center for the construction of the dock as work began on relocation or demolition of facilities and buildings in the area. Rear Admiral Frank T. Watkins, Commandant of the Thirteenth Naval District, operated the crane that dropped a 5,000-pound ball to begin crushing pavement as the symbolic start of the construction of the dock.

Manson and Osberg,[29] construction firm of Seattle, was the general contractor. Mechanical and electrical sub-contract was a joint venture by Lent's of Bremerton, Tide Bay of Tacoma and Holert of Seattle.[30]

Dry Dock 6 is 1,180 feet long and 180 feet wide and 60 feet deep with a capacity of 88,000,000[31] gallons of water. It was dedicated April 23, 1962, when tugs maneuvered the 888-foot long USS KEARSARGE (CV 33) into the dock. The 16-year-old KEARSARGE was in the Yard for modernization under the FRAM (Fleet Repair Alteration and Modernization) program.

On June 29, 1962, Rear Admiral Floyd B. Schultz became Commander of the Shipyard, and Dolan retired. With the changes of the past decade PSNS was now prepared to handle the Navy's largest ships. However, as the 1960s began, the Yard that had been the home of the Pacific Fleet's battleships and gained added prestige for its work on aircraft carriers, was about to take on a smaller, although no less challenging, type of ship — nuclear submarines.

USS BLUEBACK (SS 581) received a full SUBSAFE overhaul in 1966. Her post-overhaul inclining experiment is being conducted in superflooded Dry Dock 1.

Puget Sound Naval Shipyard

Chapter 8
The Nuclear Threshold
1962-1972

Vice Admiral Hyman G. Rickover[1] spelled out PSNS's new challenge in an address at the Bremerton Chamber of Commerce on March 27, 1964. He stated:

> *Puget Sound has a reputation for doing good work on surface ships. It was the principal West Coast battleship yard when the battleship was supreme. As battleships faded into history, the Yard took on carrier work. Now you are moving into submarine work. You cannot afford to rest on your oars — on your past reputation . . .*

Clearly, PSNS had to learn about submarines' underwater operating environment. Fortunately, the Shipyard already had a suitable facility, the Acoustic Range at Carr Inlet

Foreseeing the need for more accurate measurement of underwater acoustic noise, the Navy in 1949 had begun a search of U.S. territorial waters for a large, deep and quiet area to serve as a test range. Carl Newstrom, head of Scientific and Test Branch, recommended Carr Inlet, west of Tacoma, for the site.

Clarence A. Peterson[2] was hired to head up the eight-man Underwater Sound Section of the Design's Scientific and Test Branch. Although the range was meant to be used for research, after 1955 it began to be used more often for routine operational measurement of ship noise.

During the early 1960s, the range was used to analyze destroyers which the Shipyard had modified under the Fleet Rehabilitation and Modernization Program to establish individual acoustic signatures. Also tested were the new frigates COONTZ and KING and an icebreaker. More and more submarines ran the range and by 1965 the number of submarines tested exceeded the number of surface ships.[3]

The first regular overhaul of a submarine by the Puget Sound Naval Shipyard began in July 1962 with the arrival of the 312-foot diesel-powered CAPITAINE (AGS 336). She floated into Dry Dock 1, the same dock where the 150-foot submarine H-4 had been built in 1918. The following year, BUGARA (SS 331) received a more extensive overhaul.

This was a foretaste of Puget's role as a submarine yard. However, in the 1960s the waterfront was still busy repairing, overhauling, converting and constructing surface ships.

The workload continued to grow. The Bureau of Ships assigned lead design and construction work for a fast combat support ship, USS SACRAMENTO, to PSNS in July 1960. As in the case of IWO JIMA, SACRAMENTO was a totally new ship concept. Designed to carry stocks of boiler and aviation fuel, diesel oil, conventional ammunition, missiles, special weapons, freight and food, she provided "one-stop" underway replenishment for ships at sea.

SACRAMENTO was the first ship built in Dry Dock 4; her keel was laid June 30, 1961. In August 1962 the Yard received SACRAMENTO's main engines and reduction gears, which had been intended for USS KENTUCKY (BB 66), an IOWA-class battleship which was never completed. Each set of gears weighed more than 65,000 pounds, with overall diameters of more than 14 feet and was capable of transmitting 50,000 horsepower. These were the largest gears ever installed by PSNS.

Echoing its record breaking feats on COONTZ and KING, the Inside Machine Shop set a record in turning out Sacramento's two 56-foot long propellor tail shafts, the longest ever produced by the shop. Everything about SACRAMENTO was so huge that production and installation records were surpassed regularly.

An essential part of SAC's design was the first generation of the Fast Automatic Shuttle Transfer system (FAST), which had been developed jointly by San Francisco Naval Shipyard and PSNS. This system provided rapid underway transfer of missiles and special weapons completely separate from the handling of other cargo. This sophisticated system was a major improvement in the safe handling and rapid transfer of these sensitive weapons.[4]

SACRAMENTO was the largest ship to be built on the West Coast up to that time, as well as the largest and most powerful logistic support vessel ever constructed for the Navy. Mrs. Edmund G. Brown, wife of the Governor of California, christened the 792-foot long, 107-foot wide ship

USS CAPITAINE (AGS 336), the first submarine to be overhauled at PSNS, enters Dry Dock 1 in July 1962. Docking the submarine proved a valuable experience for later nuclear submarine dockings. The temporary guard rails on the ship's weather deck were installed to prevent workers from falling into the dock. Grosso Collection, Kitsap Historical Society

SACRAMENTO (AOE 1) in the early stages of construction in Dry Dock 4.
Puget Sound Naval Shipyard

USS SACRAMENTO (AOE 1) under way on sea trials some 28 months later provides a dramatic comparison with the construction photo.
Puget Sound Naval Shipyard

Held in place by taut lines in Dry Dock 4, SACRAMENTO (AOE 1) is ready for christening on a rainy September 14, 1963. The "herculite" strips hanging from the side shield the hot plastic anti-fouling bottom paint.

Puget Sound Naval Shipyard

on September 14, 1963, as the Thirteenth Naval District Band played "Anchors Aweigh". Captain Mark Gantar, a former ordnance officer (Code 290) at PSNS, became Commanding Officer of SACRAMENTO at the commissioning on March 14, 1964.

The success of design and construction of SACRAMENTO led to construction by PSNS of the third and fourth ships of the class: USS SEATTLE (AOE 3) and USS DETROIT (AOE 4). SEATTLE's keel was laid in Dry Dock 5 on October 1, 1965.

Traditional naval shipbuilding ceremonies emphasized SEATTLE's close association with the city across Puget Sound. A commemorative Seattle World's Fair trade dollar was inserted at the base of the mast in the Stepping of the Mast ceremony; President of the Seattle City Council Floyd Miller participated at the Boiler Light-off.[5]

Seattle Council President Floyd Miller holds the flaming torch to light off the first boiler on USS SEATTLE (AOE 3). At his right, to assist if needed, is Captain Frank J. Reh, PSNS Production Officer. Behind them are Production Department Group Superintendents James Cummins, John Longmate, Fred Timmerman and Ray "Mike" Miles.

George Carkonen/Seattle Times

Mrs. William H. Allen, wife of the president of the Boeing Company, christened SEATTLE on March 2, 1968. Seattle Mayor J. D. Braman, delivered the principal address. Braman, who spent much of his life in Bremerton, spoke warmly of Puget Sound's pride in the Navy. SEATTLE was commissioned April 5, 1969, with Congressman Thomas M. Pelly from Washington's First Congressional District, as principal speaker.

DETROIT's keel was laid in Dry Dock 4 on November 29, 1966, by Public Works Chief Engineer Paul Werner and retired Foreman Welder S. G. Thomas. Task force operations of SACRAMENTO provided information which led to improvements in the technical characteristics of DETROIT, especially in better operation of cargo handling and transfer facilities.

DETROIT (AOE 4) was christened June 21, 1969, by Mrs. Nedzi, wife of Michigan Congressional Representative Lucien N. Nedzi, who delivered the main address. Captain Robert W. Mc Clintock became DETROIT's first Commanding Officer at the commissioning on March 28, 1970.

USS CAMDEN (AOE 2), which had been built by New York Shipbuilding Corporation received her post shakedown availability at PSNS from October 16, 1967 to March 11, 1968.

While PSNS was constructing the three AOEs, it was also named lead design and construction yard for three tenders: SIMON LAKE (AS 33), SAMUEL GOMPERS (AD 37) and PUGET SOUND (AD 38). These "floating repair bases" made it possible for ships to stay in forward areas for repairs and supplies instead of returning to stateside bases.

The U. S. Navy's first built-for-the-purpose repair ship was USS MEDUSA (AR 1). USS HOLLAND (AS 1) was the first U.S. Navy tender designed specifically for the care

USS DETROIT (AOE 4) took crew, shipyard workers and family members on a seven-hour cruise around Puget Sound on August 15, 1970. Passengers and crew line the rails, eager for the ship to get under way. Puget Sound Naval Shipyard

of submarines. Both were built at the Puget Sound Navy Yard during the 1920s. Now with the submarine tender SIMON LAKE, Puget Sound Naval Shipyard would design and build a ship to tend and supply modern submarines equipped with ballistic missiles.

In contrast with diesel-electric powered USS HUNLEY (AS 31) and USS HOLLAND (AS 32), which were constructed in other yards, the steam-driven SIMON LAKE (AS 33) was a completely new design and considerably larger than HUNLEY or HOLLAND.

Captain Dennett K. Ela, PSNS Planning Officer, and Chief Design Engineer Kenneth G. Johnson attended HUNLEY's sea trials to gain background information for the design of SIMON LAKE. This new ship would be capable of supporting the new ballistic missiles and able to repair three submarines while providing logistic support for nine others.

Because of the high national priority accorded SIMON LAKE,[6] the keel-laying in Dry Dock 5 on January 7, 1963, marked the beginning of the tightest and shortest ship construction period ever scheduled at PSNS. SIMON LAKE was christened on February 8, 1964, by two daughters of her namesake, a pioneer in naval submarine development. Commissioning was November 7, 1964.

Shortly after the 640-foot long SIMON LAKE steamed off for sea trials, her Commanding Officer, Captain James B. Osborn wrote to the Shipyard Commander:

> *The plans for this ship, as developed by PSNS, represent an order-of-magnitude step forward in design of a submarine tender. The production effort has been outstanding for the ship has been built, from keel-laying to completion, in two years and actually delivered two weeks ahead of schedule. The responsiveness of your Supply department in fulfilling the many demands placed upon it by SIMON LAKE during outfitting are especially noteworthy.*

The welders in the SIMON LAKE (AS 33) keel laying ceremony from left to right are: John Jarvis, Supply Department Project Control Supervisor; Kenneth G. Johnson, Chief Design Engineer and Dean Calhoun, Structural Group Master. Among those observing are: Rear Admiral Floyd Schultz, BMTC President Paul C. Warner, NCAA President Harry F. Gundlach, NANTS President Lee O. Stafford, Chaplain Peter Bakker, Planning Officer Captain Harry Jackson, MM&FA President Colin Edwards, and Production Officer Captain J. F. Ellis. Puget Sound Naval Shipyard

On the speakers' platform, just before the ceremonial christening words are spoken for SIMON LAKE (AS 33), are: from left to right, Rear Admiral William A. Brockett, Secretary of the Navy Kenneth BeLieu, Rear Admiral Floyd Schultz, Monsignor Joseph Camerman, Senator Henry Jackson and co-sponsor Mrs. H. M. Diamond. Puget Sound Naval Shipyard

SAMUEL GOMPERS (AD 37) was the lead ship of a new class of destroyer tender designed to furnish facilites for the repair and support of destroyer-type ships, including DLGNs. Secretary of Labor William Wirtz and AFL/CIO President George Meany shared honors on July 9, 1964, in Dry Dock 4 at the keel laying of the ship named to honor one of the founders of the American Federation of Labor.

Secretary Wirtz returned in May 1966 for the christening of GOMPERS by Mrs. Joseph Holmes, Samuel Gomper's granddaughter. Captain Harry Risch Jr. took command of GOMPERS at the commissioning on July 1, 1967.

The presence of national labor leaders at GOMPERS' ceremonies was especially timely in light of Executive Order 10988, with which PSNS was the first naval shipyard to comply.[7]

On July 18, 1963, the Shipyard Commander, Rear Admiral Floyd B. Schultz, Captain W. H. Garrett and Recorder Alva E. Huguenin, as representatives of management, signed the historic agreement between PSNS and the Bremerton Metal Trades Council. Council President Paul Warner, George Giblett and Tom Landon signed for labor. Later arbitration resulted in granting Patternmakers and the American Federation of Technical Engineers separate bargaining units.[8]

Construction is almost 50 percent complete in this picture of PUGET SOUND (AD 38), shown here in Dry Dock 3. Now that the hull structure and most of the painting of the hull is finished, the forest of scaffolding shown in earlier construction pictures is mostly removed. *Puget Sound Naval Shipyard*

On September 16, 1966, as part of the celebration of the Shipyard's 75th anniversary, GOMPERS' sister ship, USS PUGET SOUND (AD 38), was christened. Captain R. B. Jacobs took command of the 643-foot long ship on April 27, 1968, in Dry Dock 6. Appropriately, the University of Puget Sound's President, Doctor R. Franklin Thompson, delivered the principal address.

Ship's company of PUGET SOUND presented the Shipyard with a large "Thank you", a 300-pound plaque with the inscription: "In appreciation for, and in commemoration of the persevering spirit and unexcelled craftsmanship exhibited in building the USS PUGET SOUND". Attached to the plaque was the welding arc stinger used by the Shipyard Commander, Rear Admiral Petrovic, for the ceremonial last weld aboard the ship. The plaque is on display in the lobby of Building 443.

PSNS also built more DLGs. Before USS KING (DLG 10) was commissioned, the keel was laid for USS REEVES (DLG 24). REEVES, a later class than COONTZ and KING was 20 feet longer with correspondingly increased breadth, displacement and armament. Her keel was laid July 1, 1960, in Dry Dock 3, where IWO JIMA was in the early stages of construction.

REEVES was christened May 12, 1962, and commissioned on May 15, 1964. Her primary distinctions over earlier DLG's were her improved fire power and the latest in submarine and aircraft detection devices.

She was followed by USS JOUETT (DLG 29) and USS STERETT (DLG 31), which received still more design improvements. When christened in Dry Dock 3 on June 30, 1964, according to SALUTE:[9]

> . . . [they had] *more freeboard forward than the COONTZ and KING and ADAMS* [of the Reeves] *class of guided missile destroyers. Another marked difference is the superstructure silhouette. While COONTZ and KING had rakish stacks and tripod masts, our latest frigates feature the combination mast and stack...*

USS REEVES (DLG 24) was moved from her building site in Dry Dock 3 to Dry Dock 2 for her christening on May 12, 1962. She is shown here moving out of Dry Dock 2 to Pier 5 where she will be completed. Larry Jacobson

USS JOUETT (DLG 29) AND USS STERETT (DLG 31) under construction in Dry Dock 3 on October 4, 1963. Dry Dock 3, which is 937 feet long was built so that it could be divided lengthwise into two watertight sections. The footing for the dividing caisson shows in the foreground of this photo. Puget Sound Naval Shipyard

USS STERETT (DLG 31), the fifth guided missile frigate built at PSNS has a full-load displacement of 7,900 tons, and is 547 feet long with a beam of 55 feet. Here on May 16, 1966, she is moored at Pier 6 nearing completion. Larry Jacobson

JOUETT was commissioned December 3, 1966 with Rear Admiral Walter H. Baumberger, Commander Cruiser Destroyer Force Pacific Fleet, as principal speaker. STERETT was commissioned April 8, 1967, Secretary of the Navy Paul H. Nitze delivering the main address.

PSNS was involved in an unusual conversion early in the decade. In 15 months it transformed the "mothballed" light aircraft carrier, USS WRIGHT (CVL 49), into a command communication ship (CC 2) with the most advanced and powerful communications systems ever installed aboard a Navy ship.

In order to eliminate electronic interference in the ship's communications, Shop 64 manufactured fiber glass masts, and used glass-impregnated roving[10] to cover metal yardarms, walkways and handrails. WRIGHT was recommissioned on May 11, 1963.

Testing and certifying the installation of complex electronic systems during WRIGHT's conversion, and the recalibration of electronics systems after submarine overhauls necessitated improvement of the PSNS Shipboard Electronic Systems Evaluation Facility. This had been established at the Ediz Hook Coast Guard Station near Port Angeles early in the 1950s.

Until buildings were obtained, William Beam, Clive Hayward and personnel from the Electronics Shop transported pickup-truck loads of equipment to Ediz Hook every time a ship went on sea trials. Harold L. Wright became Code 250.1's on-site representative, when a building on a coaling pier at the Hook was obtained for main test operations in the mid 1950s. Donald B. Buher[11] became head of the Test Facility as PSNS prepared for submarine overhauls and conversion of WRIGHT.

Buher described WRIGHT's conversion:

> *USS WRIGHT required some of the most unusual and unique electronic systems configurations used by the Navy at that time. Since the ship could become the president's command post in the event of war or nuclear attack tremendous communication system capability was needed . . . During initial planning of this project in NAVSEC* [BuShips] *it was often jokingly referred to as "Project Nightmare".*

This aerial view of April 1963, ranging from Sinclair Inlet to the Olympic Mountains, shows a busy waterfront with USS WRIGHT (CC 2) at Pier 4 and USS TICONDEROGA (CV 14) in Dry Dock 5. Also in the picture are: USS WINONA (WHEC 65), USS EVERSOLE (DD 789), USS ARNOLD J. ISBELL (DD 869), USS BRINKLEY BASS (DD 887), USS COLUMBUS (CG 12), USS BRADFORD (DD 545) AND USS BROWN (DD 546). Larry Jacobson

The technological advancements that the Navy incorporated into its ships also made it possible to explore the ocean at depths previously impossible to reach. In addition to scientific gains, this gave the Navy a means of rescuing submarine personnel from greater depths than ever before.

PSNS, involved in this part of the Navy's deep submergence projects, designed changes in submarines to facilitate use with Deep Submergence Rescue Vessels, (DSRV). These small but completely operational submarines could be flown to the rescue site or fastened to a "mother" submarine and taken piggy-back to the rescue site. Near the disabled sub the DSRV would move on its own power to link up with the larger submarine and rescue its personnel.[12]

After a buildup in 1962 because of the Cuban crisis, the PSNS workforce dropped to less than 8,900 in January 1964. However, because of U. S. involvement in Vietnam, the work force began a new rise before the end of that year to reach a high of 10,850 in August 1967.

In addition to its other work, PSNS activated LCUs and YTLs in 1965. Two years later, the Shipyard commissioned and converted two former Army supply ships into electronic surveillance ships. These were USS PUEBLO (AGER 2) and USS PALM BEACH (AGER 3). Following a five-month availability at PSNS, NUECES (APB 40) was recommissioned on May 3, 1968, as a self-propelled barracks ship for an Army battalion. Painted green for camouflage, NUECES left to join two other APBs on Vietnamese rivers, for use as a forward supply base and field hospital.

Many changes were taking place in the Shipyard. Of prime importance was expansion of the Yard's facilities and the increased training necessary for PSNS to become a nuclear submarine repair facility. The effort began in 1955 when Shipyard Commander, Rear Admiral Dolan, authorized the Industrial Relations Training Division to prepare a long-range nuclear power training program.[13]

Commander Archie P. Kelly[14] was assigned to PSNS to assist in the development of the program. Joseph L. Lambert and William C. Spiller went to Washington, D. C., for training under Rickover from November 1955 to January 1958.

In September 1961, Planning Officer Ela, told SALUTE that Design Engineers Raymond L. Kronquist, John Lansberry, Jack H. Nuszbaum, John D. Prebula, Spiller and Lambert[15] were the nucleus of the PSNS' Nuclear Power Training Unit which would become the Nuclear Power Division. He added that a Navy officer with a billet title of

Deep Submergence Rescue Vessel TRIESTE II, shown here at the Naval Undersea Warfare Engineering Station, Keyport, is representative of the vessels on which the Yard made design modifications. *Naval Undersea Warfare Engineering Station*

USS BARBEL (SS 580), the Yard's third submarine overhaul, was the first submarine to enter the SUBSAFE program at PSNS and to be certified by the Bureau of Ships. Grosso Collection, Kitsap County Historical Society

Nuclear Power Superintendent, would head the division.

This would be Commander William Wegner. Early in the training period he took the Production Department Group Superintendents to Idaho Falls and Arco, Idaho, so they could become familiar with nuclear prototypes. Captain William S. Humphrey relieved Wegner as Nuclear Power Superintendent in the fall of 1962 and began developing the facilities necessary for working on all types of nuclear powered ships.

The Shipyard adopted new working restraints, necessary because of the use of nuclear power. The Medical Department's mission was changed to support the program; new personnel were hired for the Industrial Hygiene Division. Classes were held in radiological safety and a new radiation detection film badge program was initiated. All departments in the Shipyard were assigned Nuclear Coordinators.

The first nuclear-powered ship to visit PSNS was USS HALIBUT (SSN 587) in February 1960. Her stay was brief, merely to give the crew week-end liberty before proceeding to the Carr Inlet Acoustic Range. Her welcome was hearty. The ship's captain, Lieutenant Commander Walter Dedrick, said, "We never expected anything like this," while the crew, lining the deck, responded to the enthusiastic welcome from the Shipyard and Bremerton.

In April 1964, PSNS began its third overhaul of a diesel-powered submarine. This was USS BARBEL (SS 580). The overhaul requirements included a new submarine safety program then being directed by the Bureau of Ships.

Following the loss of USS THRESHER in April 1963, all U.S. submarines were restricted in diving depth until certified as being safe to operate at the designed operational depth. It was necessary for PSNS to become qualified in the new certification program, familiarly known as SUBSAFE.

A group of Design Division engineers, headed by Bruce Lent, went to the sub base at Pearl Harbor for the pre-overhaul ship check on BARBEL; another group, headed by George Campbell,[16] reported to the Charleston Naval Shipyard to become familiar with submarines and their components. They began the SUBSAFE programs for those two Shipyards before returning to PSNS.

USS SNOOK (SSN 592) was the first nuclear submarine overhaul at PSNS that included refueling. This fifteen-month availability was made easier on the crew by changing the ship's home port to Bremerton for the overhaul period.

Puget Sound Naval Shipyard

SUBSAFE involves preparing "mapping drawings" which locate and identify all hull penetrations, piping components and piping joints that required certification and the method of certification to be provided.

According to Campbell:

> *Accomplishing these jobs became a herculean task that involved every shop and office in the Shipyard. Special organizations were formed so that final inspections and certification records could be assured and audited for final certification. This and the regular overhauls of submarines continued to be the dominant task for the Shipyard . . .*

Emphasis on quality assurance and control was strong. A "clean room" for work on submarine hydraulic valves and components was set up in Building 431, to establish and maintain a cleanliness level for these items. "Grade A" levels of cleanliness became a familiar term throughout the Shipyard.

To meet the required measuring standards, the Paint Shop procured a gauge to determine the thickness of paint coats applied to hulls and interiors of ballast tanks. PSNS developed a new method to measure a hull's circularity precisely, a vital factor when patching huge access holes required for removal and reinstallation of shipboard components.

Each submarine received an emergency ballast tank blow system to allow the submarine to surface, after a flooding, independent of its standard operating systems.[17]

With BARBEL, PSNS became the first naval shipyard to complete a SUBSAFE overhaul with results approved by BUSHIP representatives. PSNS completed a full SUBSAFE overhaul in 1966 on USS BLUEBACK (SS 581), ahead of schedule.

The Yard completed its first regular availability of a

After construction at New York Shipbuilding Company in Camden, New Jersey, USS TRUXTUN (DLGN 35) came to PSNS for her post shakedown availabilities. She arrived here on September 1, 1967 and was the Yard's introduction to work on nuclear-powered surface ships.

Larry Jacobson

nuclear powered submarine — USS PLUNGER (SSN 595), in March 1968. The first nuclear refueling availability, USS SNOOK (SSN 592), was completed three months later.

An organizational change took place the fall of 1968. Rickover told Petrovic, that all work and responsibility for ship nuclear work, would be turned over to individual naval shipyards within two years. Not wasting any time, Petrovic arranged for Commander James H. Webber and Lieutenant Commander Roger B. Horne to be assigned to the Yard.[18] They arrived in August 1968.

Webber became the Shipyard's Nuclear Power Superintendent, with Horne as his assistant. Humphrey headed the new Naval Reactors Representative's Office with responsibility for oversight of all nuclear work and reported directly to Rickover.[19]

Webber and Horne set up a school to train Engineering Duty Officers as nuclear-ship superintendents, a move copied in other naval shipyards. Horne also assumed a collateral duty as Nuclear Repair Superintendent in the Production Department. PSNS conducted schools for Production Department personnel, which were also attended by officers and civilians from other Naval Shipyards. When Webber became the Yard's planning officer in 1970, Horne became Nuclear Power Superintendent.

PSNS's first overhaul of a ballistic missile submarine began in August 1968 on USS JOHN ADAMS (SSBN 620), the Yard's most extensive submarine overhaul to that date. Commander Edward Mortimer was Commanding Officer of the ship's Blue Crew.[20] More ballistic missile submarine overhauls, some including refueling, followed in rapid succession: U. S. GRANT, PATRICK HENRY, ROBERT E. LEE, LEWIS and CLARK, GEORGE MARSHALL and FRANCIS SCOTT KEY.

USS TRUXTUN (DLGN 35) arrived September 1, 1967, to be PSNS's first introduction to work on nuclear powered surface ships. She was followed by USS ENTERPRISE

Sea water damage of USS CONSTELLATION (CVA 64)'s propulsion shafting provided the Yard a new challenge. The upper photo shows the severe erosion near a coupling flange on an intermediate shaft. The lower photo shows the repaired shaft ready for re-installation. Puget Sound Naval Shipyard

(CVAN 65) in July 1968. TRUXTUN returned in January 1969 for an intense three-month availability that placed heavy demands on the Pipe Shop.

The Shipyard's involvement in nuclear work required additional security because of the sensitive nature of the work. In 1963 the Yard issued bumper stickers to replace the former windshield decals and decreased the number of vehicle passes issued.

The most visible sign of increased security was an eight-foot, barbed-wire topped, steel-link chain fence around the Controlled Industrial Area (CIA) in 1966. Almost a mile in length, the fence had new gates and guard houses at the State Avenue entrance and where the fence crossed Farragut Avenue in the west end. USS MISSOURI was moved to a berth at the west end of the Yard, outside the CIA, to permit continued visits by the public.

Carriers continued to come in for overhaul or repair work. USS CONSTELLATION (CVA 64) provided a particularly interesting job for PSNS. When she arrived in June 1970, inspection revealed erosion of unprecedented severity on her main propulsion shafting. Protective covering material had carried away, exposing couplings to salt water; the bolts were badly eroded. In some areas of the shaft the depth of the erosion was as much as an inch, with possible serious consequences.

Because of the high cost of replacing the shafts, the decision was made to repair them, a decision influenced by the advanced "state of the art" welding technology and facilities at PSNS. Essentially all work was done within the Machine Shop.

This required careful coordination and scheduling of various phases of preparation and repair, such as sandblasting, heat-treating, machining, testing and preservation. Local area weld repairs were accomplished using the "shielded metal arc welding" process. Restoration of metal to large eroded areas was accomplished by pre-heating the shaft region to be welded and then depositing weld by the "gas metal arc welding process".

After shaft rebuilding was completed, the shafts were sized on metal turning lathes and rubber coated for protection against further corrosion. The complete job of removal, repair and realignment of the shaft was accomplished in less than six months, allowing the ship to maintain its operating schedule.

In a report dated February 16, 1971, Inside Machine Shop Superintendent Mel D. Wortman and Welding Shop Superintendent Harold Schroy summarized:

> *The shaft repair work involved approximately 32,000 man hours of work to apply 13,500 pounds of welding rod, to replace approximately 60,000 cubic inches of material. Approximately 5,000 pounds of new bolts were forged and machined to fasten together 920,000 pounds of shafting that required over 6,000 pounds of rubber for protective cover.* [21]

A new type of ship arrived in July 1963. This was the hydrofoil, HIGH POINT (PCH 1) which, with PLAINVIEW (AGEH 1), became the Hydrofoil Special Trials Unit at PSNS.[22] The unit conducted sea trials, evaluating different design configurations and equipment on hydrofoils.

PLAINVIEW was docked in Dry Dock 1, which in 1967 had been altered to permit superflooding. Parker Snapp called it:

> *. . . a neat trick that the Shipyard learned when we started to handle submarines which draw too much water even when in a light condition, to float into our drydocks and clear the keel blocks. They float in alongside the blocks, then the caisson is closed and the interior superflooded until the submarine has clearance to slide sideways over the blocks. Vulnerable galleries at the old highwater mark had to be walled up with dikes before superflooding could be conducted.*[23]

USS PLAINVIEW (AGEH 1) at the crest of super flooding in Dry Dock 1, when the water level was less than one foot below street level. Super flooding was necessary to dock PLAINVIEW because she had to be docked on eight-foot-high blocks to permit her foils to be lowered.

Puget Sound Naval Shipyard

A plaque commemorating the first flag raising at PSNS was dedicated on September 16, 1966. Here offering the invocation for that ceremony is Monsignor Joseph E. Camerman, pastor of Star of the Sea Church in Bremerton. The lower photo shows the plaque. Puget Sound Naval Shipyard

A major event during the Yard's seventy-fifth anniversary celebration was the salmon barbecue prepared by the supervisors. These beautiful slabs of salmon are being watched over by Anniversary Chairman Stan Clarke, Pipe Shop Foreman Curtis Bay and accordionist Myron Floren. Grosso Collection, Kitsap County Historical Society

The Puget Sound Naval Shipyard was 75 years old on September 16, 1966. In preparation for its diamond jubilee celebration, points of interest in the Yard were marked with bronze plaques similar to the one placed at the head of Dry Dock 6 when it was dedicated. Plaques were placed near the other five dry docks, the tunnel below the officers' quarters, Building 50, the Machine Shop, Shipfitters Shop, and Apprentice School. Two plaques were put in front of Quarters C, calling attention to the first five quarters built and the Giant Sequoia tree standing in front of Quarters C.[24]

Stanley D. Clarke of the Industrial Relations Office was chairman of the anniversary committee. Tickets admitting anyone over 6 years of age were $1.50. Over 10,000 people attended the two-day celebration.

Friday, September 16, was a full day, starting with the plaque dedication at Dry Dock 1,[25] at which a retired former Shipyard Commander, Vice Admiral Homer N. Wallin, was the principal speaker. This was followed by a luncheon in Building 435 for over 400 "old timers", those whose employment in the Yard began before September 16, 1921. At 5:30 Mrs. McGee, wife of Senator Gale W. McGee of Wyoming, christened USS PUGET SOUND in Dry Dock 3. That evening there was a dance for teenagers in Building 502.

Saturday started with a salmon bake in the west end of the Shipyard, followed by entertainment. Three shows were given by accordionist Myron Floren and dancers Barbara Boylan and Bobby Burgess, all of whom were members of Lawrence Welk's Champagne Music group. Building 502 was the site of the anniversary ball that night.

New equipment was acquired to handle the current and future workloads. During 1967 and 1968, PSNS received 16 cranes and other equipment from de-activated New York Naval Shipyard.

The large size thick wall piping to be installed in the new ships exceeded the fabrication capability of Shop 56 equipment. In February 1962, a high performance pipe bender was acquired from machine designer and manufacturer John L. Costello. Here, on the right side of the picture, observing the demonstation are: Captain R. J. Nesbitt, Production Officer, Rear Admiral William Dolan, (mostly hidden) and Shop Master Colin Edwards. Puget Sound Naval Shipyard

On February 2, 1962, SALUTE reported the arrival of the Pipe Shop's new 16-inch pipe bending machine:

> *Shipyard management . . . this week was treated to a preview of the productive capabilities of Shop 56's giant new radial bending machine, acclaimed the largest of its kind now operating in these United States. Hydraulically operated, the bending machine can handle extra heavy steel piping upwards of 14-inches in diameter and with a wall thickness of a half-inch with pretzel bending ease . . .*
>
> *Considered a conservative estimate, a neat savings of $75,000 is expected alone in the bending of AOE-1 piping, . . . Presently known AOE-1 piping requirements, in the seven through 20 inch in diameter range, affecting forthcoming use of the new bending machine include steel piping and tubing, copper tubing and copper-nickel alloy piping totaling 12,651 lineal feet. This figure, according to Master Pipefitter* [Colin] *Edwards, already establishes AOE-1 as the biggest single ship piping job ever undertaken by Shop 56 . . .*

The largest vertical boring mill in the Northwest was ready for operation in the Machine Shop in October 1966. Ten months earlier Public Works had begun excavation in the north end of Building 431's center bay for pouring the largest single foundation for a machine in the Shipyard's history.

Into the base, workers sank a 3x3x18-inch deep metal box containing rosters of all Shop 31 personnel and those of Public Works and Shop 72 who were involved in preparation of the installation of the mill. It also contained sample order forms, work schedules, a PSNS phone directory, pictures and newspapers. The cover of the time capsule carried the instructions: "Remove when making room for more modern machinery, 3-21-66".[26]

The new machine, capable of handling pieces up to 185,000 pounds, replaced the mill installed in 1914, which could not handle anything over 35,000 pounds. Jubilant Foreman A. J. (Archie) Merrifield told SALUTE:[27]

> *The new 30-foot vertical boring mill, once in operation, will most certainly put us in the enviable position of no longer having to reject, or farm out machining work of the magnitude this new machine will handle.*

In 1960, PSNS received one of Navy's first numerically controlled machine tools, a multiple-spindle drilling and

An early version of an electronically controlled metal machining tool in Shop 31. Receiving input from a numerical paper tape, this turret drill makes identical reproductions of the hole size, angle and spacing on the work piece. Additional support brackets in the foreground await work.

Puget Sound Naval Shipyard

The electronically controlled "Telerex" machine cuts precisely dimensioned plates for ship construction. The two torches produce identical plates or mirror images, controlled by paper tape numerical input or photo electric tracing. Watching an early test run are Production Officer, Captain J. F. Ellis, Group Superintendent Dean Calhoun and others.

Puget Sound Naval Shipyard

tapping machine capable of operating from coded information punched in a paper tape. Enthusiasm for this radically new machine was initially low because it was considered suitable only for large quantities of repeat type parts.

However, before the end of the decade, Shop 31's machinists were skilled at the operation of electronically controlled consoles, while tools faced, drilled, milled and finished the work. In Shop 51, using the new automatic wire stripping machine, electricians cut and stripped wires in less than 20 percent of the time required by hand. In addition to the savings in time, the machines assured exact duplication impossible by even the most skilled mechanic, a factor increasingly important because of the Yard's nuclear work.

A change was made in the duplication of paper plans and reports. Oliver M. Vanderkleed, who worked in Planning's Reproduction Section from 1937 to 1969, called the Xerox 1860:

One of the most useful and significant advancements and valuable pieces of equipment added to our shop . . . This marvelous piece of equipment copied original drawings and other documents and automatically reduced them to one-half size . . . We made the half-size prints on a reproducible vellum and from these printed one-half size diazo prints for the shops and offices. Prints were reproduced on a very high speed Revolute diazo printer at speeds up to 60 linear feet per minute. The use of blueprint paper was greatly reduced.

SALUTE in June 1961 reported the expected arrival of the copier, noting the savings to be expected and the great convenience for "workers who have had more than a few words to say about the unfolding, flapping, unhandy size of blueprints on the production line." The article explained that "half-size" referred to both length and width so the prints would actually be only one-quarter size and cost 15 cents, against 50 cents for the full size print.

By 1969, the Shipyard's data processing operation and

Originally built as a general warehouse, Building 467 became the headquarters building for the Naval Supply Center Puget Sound, when it was established in October 1967. Some of the interior spaces were remodeled to become offices for the new command.

Puget Sound Naval Shipyard

personnel in Building 467 had been separated from the Comptroller Department and set up as the Data Processing Office, Code 110, headed by Stephen P. Cartwright, who was also head of the Yard's Industrial Management Office, Code 140.[28]

In 1970 the Data Processing Office effected considerable savings by buying additional equipment instead of leasing it. In addition to providing services to the Shipyard, the Office scheduled and operated the Naval Supply Center's computer system.

The Naval Supply Center Puget Sound was officially established on October 2, 1967, with the consolidation of the Naval Supply Depot Seattle and most functions of the Shipyard's Supply Department. Approximately 34.4 acres of land at the west end of the Shipyard were transferred to NSCPS; Building 467 was remodeled to provide offices for the new command.

NSCPS assumed all Naval Supply System support functions. The Shipyard Supply Department retained responsibility for industrial material support, shop store inventories, and direct material inventories. The Supply Department transferred 514 civilian and 10 officer billets to the Supply Center, retaining 244 and 6 billets respectively.[29]

Administration of the Naval Fuel Depot at Manchester was transferred from Naval Supply Depot Seattle to NSCPS. This beautiful property was obtained from the Army Department[30] by the Navy Department in 1924. Limited use was made of it until 1939, when the Navy began development of the Fuel Storage Depot to support fleet and shore units during World War II. During the Korean War additional facilities were constructed.

At the former Naval Ammunition Depot at Ostrich Bay, approximately 133 acres were transferred to Bureau of Yards and Docks in 1965 for housing construction. This was the beginning of Jackson Park Housing. U. S. Senator Henry M. "Scoop" Jackson dedicated the complex of 100 homes on June 13, 1966. Adjacent land was transferred to the City of Bremerton and the Central Kitsap School District.[31]

To clear land for the construction of an Enlisted Men's

Part of the land that had been the Bremerton Naval Ammunition Depot was used to provide more Navy housing. This first increment of Jackson Park, as the housing area was named, provided housing for 100 families and was dedicated by Senator Henry Jackson on June 13, 1966.

Kitsap County Historical Society

Ground breaking for the new BOQ on the hill above the Officers' Club took place in September 1968. Wielding the shovels, from left to right, are: Captain Ralph C. Jensen, Public Works Officer; Captain John W. Gardinier, Officer-in-Charge-of-Construction and Rear Admiral Petrovic, Shipyard Commander.

Puget Sound Naval Shipyard

Club,[32] the Shipyard scheduled the demolition of Craven Center. The Puget Sound Naval Shipyard Museum in the Center closed December 30, 1966, and the following year was moved to the Washington State Ferry Building.[33] General Services Administration conveyed title of the remainder of the former Navy Yard Hotel site to the City of Bremerton. The land is now used for two parking lots and the continuation of Park Avenue between Fourth Street and Burwell Street.

There were new buildings in the Yard itself. On the hill above the officers' quarters a new five-story Bachelor Officers' Quarters, Berkheimer Hall, was completed early in 1970. The foundation of the Shipyard's new Engineering and Management Building (Bldg. 850), was built into the hillside below the officers' quarters. Senator Jackson gave the principal address at the dedication on July 2, 1970. The facility almost doubled in size in 1981, with the addition of Building 850A.

In the west end of the yard a new Marine Barracks (Bldg. 853) was dedicated on April 30, 1971. This replaced the barracks built in 1910 which had been damaged and declared unsafe after an earthquake on April 29, 1965.

Heavy snow late in January 1972 made it difficult for workers to reach the Yard. One incident was recorded humorously in SALUTE on February 4.

Marines at the Farragut Avenue gate were astounded when two horseback riders appeared. Chief Engineman Harry Starr and his son had ridden their quarterhorses from their home on the Brownsville Highway so the chief could get to his job at the Naval Inactive Ship Maintenance Facility,[34] tenanted inside the PSNS fence. The Marines ascertained that horses could be admitted to the compound. Starr reported for duty and received a "Well done" for his ingenuity from Captain Alan Dougall, his commanding officer.

PSNS civilians and military continued to receive their share of awards. Supervisory Industrial Hygienist Daniel J. Bessmer was named Outstanding Federal Employee of the

Senator Henry M. Jackson looks at the plaque unveiled in front of Building 850, the new Engineering and Management Building which he dedicated on July 2, 1970. The building itself is shown in the lower photo. Building 850 houses offices previously in Buildings 78 and 104.

Puget Sound Naval Shipyard

The Navy Band plays the National Anthem, the Marine Honor Guard "Presents Arms" and other military personnel salute as the flag is raised at the dedication of the Marine Barracks (Bldg. 853) on April 30, 1971. Puget Sound Naval Shipyard

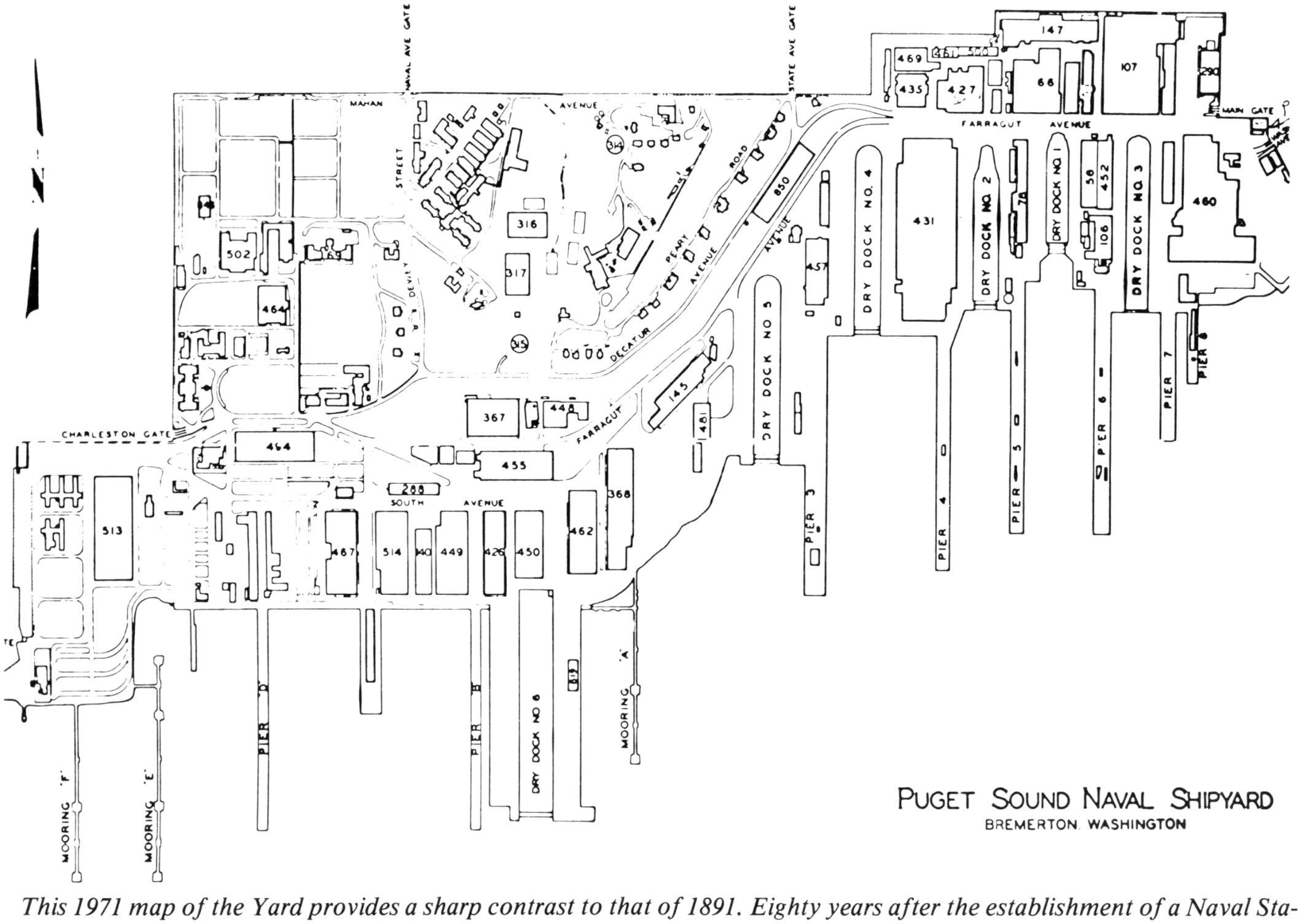

This 1971 map of the Yard provides a sharp contrast to that of 1891. Eighty years after the establishment of a Naval Station centered around one dry dock, there is very little land not in use. Puget Sound Naval Shipyard

An unusually heavy snow fall showing Quarters "E" and the duplex Quarters "F" and "G" is recorded in this photo taken from the new Bachelor Officers' Quarters. The empty waterfront results from the workload of submarines, most of which are hidden from visibility in the dry docks.
Puget Sound Naval Shipyard

Year in 1968 by the Puget Sound Federal Executive Board.

Bessmer was chosen for his nationally known accomplishments in industrial health. That same year he received the J. M. Dalla Valle Award presented by the American Industrial Hygiene Association. Bessmer led the Shipyard's Industrial Hygiene program under the pressure of the industrial technological explosion that followed World War II, with the increasing complexity of methods and materials used.

The early years were spent finding protection for the workers' eyes and lungs; safety shoes became mandatory. Applications for hearing loss compensation in 1967 were reduced 52 per cent from 1966, following the requirement that ear protection be worn in noise hazardous areas.

The PSNS Industrial Hygiene Division in 1970 developed a suit to protect painters from dermatitis caused by epoxy paint spray. That same year the Division published a technical manual on controlling exposure to asbestos. Leaders in the field visited PSNS and Doctor Duncan Holaday of the U.S. Health Service told SALUTE:

> *In many places, some work has been done towards reducing the exposure for man, but Puget Sound Naval Shipyard has gone the farthest in this line and developed the most satisfactory operating program which we have seen to date.*[35]

Bessmer retired in July 1972, a year that saw many retirements. With the withdrawal of our troops from Vietnam and the post-war reductions in the armed forces, the Navy dropped the employment ceiling at PSNS to 8,100. Fortunately, because of the large number of employees taking advantage of "early out" retirements, relatively few employees were laid off.

The Shipyard Commander also retired. Petrovic turned command of PSNS over to Captain Francis F. Manganaro on June 30, 1972, in traditional ceremonies held for the first time on the roof of Building 850.[36]

Prior to the change of command, Petrovic had officially accepted for the Shipyard, the new woodworking plant (Bldg. 851), the first increment of the Shipyard's facility modernization of the next decade.

A trunk section of a 165-foot-high Douglas fir tree from Jackson Park became the focus of the Shipyard's memorial for the National Bicentennial. The plaque at the left of the log identifies the national historic events with the numbers marked on the tree's growth rings. The plaque at the right commemorates the Centennial Tree, and identifies its species and source.

Donald Serry

Chapter 9

The Changing Skyline

1972-1981

In the mid-1960s the Bureau of Ships commissioned Kaiser Engineering to provide a long range modernization plan for all naval shipyards. The first visible sign of this program at PSNS was Building 850, the Engineering and Management Building, dedicated in 1970. This building was included in the Kaiser plan, although funding had been authorized earlier, because of earthquake damage in Building 78.

During the early 1970s, Kaiser's "Engineered Long Range Modernization Program" resulted in major changes in PSNS' capabilities and appearance. New buildings for the Woodworking and Sheetmetal Shops and a Nuclear Repair Facility created a new Farragut skyline as buildings were raised or demolished.

Work on Shop 64's new Building 851 began in 1972. Its 40-ton bridge crane in the boat shop bay and a 30-ton crane in the woodworking bay were successfully tested in February

The 84,675 square feet of work space in the new Woodworking Shop, Building 851, became the home for Shop 64 in 1972. Building 66, the Shop's former work place, was taken down shortly after the Shop moved to Building 851.

Puget Sound Naval Shipyard

The area of a major change on the north side of Farragut Avenue outlined in white, delineates Building 66 and the Parkerizing Plant, Building 418, which were taken down to make room for the Nuclear Repair Shop. Venerable Building 59, the Pattern Shop, remains.
Puget Sound Naval Shipyard

1972. The new $2,696,585 building, constructed between Dry Docks 5 and 6, was dedicated on May 20, 1972. It provides 84,675 square feet of floor space.

With the completion of Building 851, the former woodworking shop complex, Buildings 66, 108 and 109, was razed. Building 66 had been built in 1900 as the Steam Engineering Building and used as the Machine Shop until 1935. A huge chart recording PSNS' World War II ship building and repair accomplishments had hung on Building 66's front exterior until the late 1960s.

The Nuclear Repair Facility was completed in March 1973 on the site of Building 66. Constructed of structural steel and precast concrete panels, Building 856 is across Farragut Avenue from Dry Dock 1.

Nearer the Main Gate, the Sheetmetal Shop's building was modernized and doubled in size in a two phase program. The first part of Building 857 was completed in November 1972, adjacent to the shop's Building 102; the shop's equipment was moved into the new building. Upon completion of this phase of construction, Building 102 was razed and the second part of the new building constructed in its place. Completed in the spring of 1974, Building 857 is constructed of precast concrete panels. It fronts on 160 feet of Farragut Avenue and extends 500 feet to the north.

Farther west, on the south side of Farragut Avenue, a two-story warehouse (Bldg. 145) came down to make room for a new 38,000-square-foot Metal Surface Preparation Facility, Building 873. This building provides a single location for sand blasting, painting, chemical cleaning, plating and other metal preparation operations. With all such operations in one building, time is saved and pollution is minimized. It is easier to collect and contain air-borne par-

When Building 66, which had housed the Boat and Joiner Shop, tumbled to the wrecking ball in 1972, over 70 years of memories fell in the ruins. Building 66, built in 1899, was enlarged by several additions over the decades. The Nuclear Repair Facility now occupies this site.

Puget Sound Naval Shipyard

ticles and surface treatment chemicals when work does not have to be transported to other buildings.

Late in the summer of 1975 the Shipyard gained a newly remodeled police station on First Street at Pacific Avenue. Building 497 had been acquired from the Bremer Estate in 1942 and renovated to house the police and marine guard. In 1975 the police area was updated to include a court room and a reception center built in the space vacated by the shore patrol, who moved into Building 433. The outside received new paint and re-designed windows. Functions, such as display of job openings and receipt of Form 171 applications, which had been performed at the foot of the Burwell Street employment ramp in Building 290, moved to the new reception center.

Other changes were made in the industrial area. Building 858, the Plate Cutting Shop, was erected east of Building 460, the Shipfitters Shop. Its Telerex machine, with digital tape inputs, automatically cuts metal sheets in a fraction of the time required when the work was done on the Shipfitter's Slab.

Building 78, built in 1903 as the Equipment Building but used from 1910 to 1970 as the Administration Building, remained in its place on Farragut Avenue; Building 104, attached to the south end of Building 78 in 1917 by a three story addition, was razed. A Waterfront Support Building (Bldg. 879) was constructed on the site of Building 104.

Two new programs tackled the problem of air-pollution abatement. "Bag-houses" in the Foundry, Pipe and Boiler Shops collected and disposed of smoke, fumes and dust particles. Two electronically operated incinerators, capable

Visible on the left, looking east on Farragut Avenue in February 1973, are the new Nuclear Repair Facility on the site of the old Woodworking Shop (Bldg. 66), the Pattern Shop, Boiler and Pipe Shops. In the distance, beyond the head of Dry Dock 1, right foreground, is the Shipfitters Shop.

Puget Sound Naval Shipyard

The Sheetmetal Shop, constructed in two sections in 1971-1972, replaced Building 102, which had been built in 1905 for the same shop. Building 857, which houses a modern sheetmetal facility, provides 82,000 square feet of working floor space.

Puget Sound Naval Shipyard

Windows of the red brick YMCA (upper left) look down on the extensively remodeled Shipyard Police Station in July 1975. The Shipyard acquired the Hensel Building which housed the Olympic Tavern, from the Bremer Estate in 1942. Lieutenant Ambrose B. Wyckoff Park is at the right. Puget Sound Naval Shipyard

Carriers KITTY HAWK, RANGER and ORISKANY, AOEs SACRAMENTO and CAMDEN and the cruiser BAINBRIDGE are visible in this February 1977 aerial view. New Shipyard buildings are labeled: (1) Woodworking Shop, (2) Metal Preparation Building, (3) Engineering and Management Building, (4) Nuclear Repair Facility, and (5) Sheetmetal Shop. Puget Sound Naval Shipyard

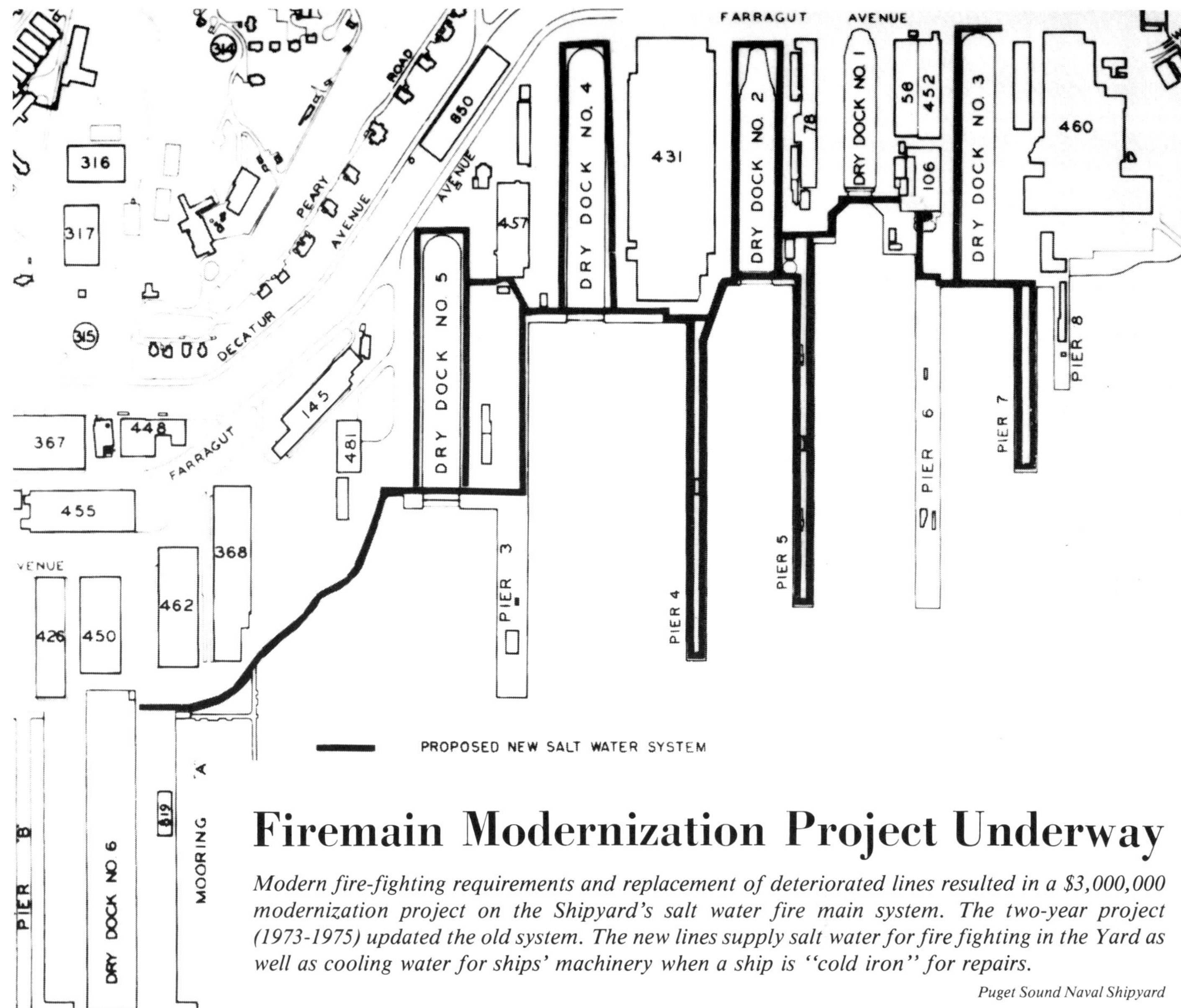

Firemain Modernization Project Underway

Modern fire-fighting requirements and replacement of deteriorated lines resulted in a $3,000,000 modernization project on the Shipyard's salt water fire main system. The two-year project (1973-1975) updated the old system. The new lines supply salt water for fire fighting in the Yard as well as cooling water for ships' machinery when a ship is "cold iron" for repairs.

Puget Sound Naval Shipyard

of burning 350 pounds of paper an hour with virtually no smoke or odor, were installed in Building 550 in the southwest corner of the Yard.

Work was begun in 1969 to separate the Shipyard's storm and sanitary sewer piping systems. This required extensive excavation throughout the Yard, but reduced the amount of water flowing into the City of Bremerton's waste-water treatment plant.

PSNS modernized its salt-water fire main system during the early 1970s. Fire Chief R. E. Fletcher and his crews welcomed the new 20-inch main line, which increased the system's capacity from 350 gallons per minute at 50 pounds per square inch to 7,000 gallons per minute at 150 pounds per square inch.

The Quality Assurance Department's Metrology Laboratory[1] went underground in October 1973, from Building 435 into the World War II air raid tunnel in the hill below the officers' quarters. The move was necessary, in part, because interference from a power distribution sub-station being built next to Building 435 would interfere with operation of the sensitive calibration equipment.

The laboratory moved into 330 feet of the main tunnel and 145 feet of Portal 3.[2] The eastern end of the tunnel was already occupied by the Telephone Exchange which had moved there from Building 78 in August 1970 when the new Central Exchange System became operational.[3]

The PSNS Metal Trades Council Credit Union moved into its newly constructed building at 1025 Burwell Street in January 1974. This was the first time the Credit Union's offices had been outside the Yard since 1947, the year in which membership had been extended to all Shipyard employees and an office set up in Building 290.

Former Chief Planner Lloyd A. Prichard and former Draftsman H. W. Schwartz attended the dedication ceremonies of the new building. Forty years earlier they and five other employees of the Puget Sound Navy Yard signed an

While torrential rains poured down on July 7, 1976, four naval officers and two political officials dug their shovels into a wooden crate of black soil to symbolize the ground breaking for the new naval hospital. From left to right: Senator Henry M. Jackson; Rear Admiral H.A. Sparks, Commanding Officer of the Naval Regional Medical Center, Oakland; Captain H.P. Pariser, Commanding Officer of the Naval Regional Medical Center, Bremerton; Rear Admiral J.D. Murray, Thirteenth Naval District Commandant; Captain C.M. Howe, Commanding Officer of the Western Division, Naval Facilities Command, and Congressman Floyd V. Hicks. The ceremony took place in a Jackson Park recreation building not far from the hospital site.

Kitsap County Historical Society

application for a charter for a credit union.[4] The Credit Union started with assets of $35 and 18 members of the Metal Trades Union Council, who held their meetings in private homes.

A new name, Kitsap Federal Employees Credit Union, was chosen in April 1981, to reflect its membership, which now included employees of other federal installations as well as military personnel. The Credit Union's sixth branch opened in 1988, and once again the Credit Union operated within the fence, this time in the northwest corner of the Shipyard.

Two tenant commands were separated from PSNS during this period. On July 7, 1976, ground was broken for a new naval hospital north of the Jackson Park Housing area. The nine-story, $22 million dollar Naval Regional Medical Center opened in April 1980, staffed by 50 doctors, 50 commissioned nurses and 200 hospital corpsmen. The building contains a variety of energy-conservation features including heat recovery equipment and its own steam engineering plant.[5]

The three brick hospital buildings which opened in 1911 in the Puget Sound Navy Yard were demolished in 1981, as

This 1951 photo shows the Marine Officers' Quarters, parade field and Barracks in the upper left corner. The lower two-thirds of the photo shows the Naval Hospital complex, most of which was demolished in 1981. Kitsap County Historical Society

The new Naval Hospital, erected near Jackson Park, is a modern, well equipped nine-story building which began operations in 1980. The photo shows the main entrance and the southern elevator shaft. Consultation and treatment areas are in the lower three floors.
Kitsap County Historical Society

These two log buildings were built by Civilian Conservation Corps at the Ammunition Depot on Ostrich Bay in 1935. They were transported by barge across Dyes Inlet to the Silverdale waterfront, adjacent to the Kitsap County Historical Museum. Originally the house was the home of the Depot gardener, and the stables were used by Marine guards stationed at the Depot.

Puget Sound Naval Shipyard

The nation's Bicentennial in 1976 was observed in the Shipyard with the development of a triple memorial. A section of a 200-year-old tree from the Ammunition Depot was marked with significant dates; a hillside of red, white and blue rhododendrons was planted in the shape of a pennant and a memorial marker was enhanced by an archway under which was suspended the Bicentennial logo surrounded by a bell-shaped length of chain. The Shipyard Bicentennial Committee members (insert alongside the arch) are Harold O. Pickard and Donald Serry. Donald Serry

were most of the brick and frame structures that had been added through the years. Exceptions were the hospital officers' quarters and Buildings 491 and 443 which became offices, and Building 437, the present "Outfitters."

The Marine Barracks also moved. As part of a nationwide effort to simplify Marine Barracks operations, the Bremerton Marine Barracks was combined administratively with the Barracks at the Naval Torpedo Station Keyport in December 1976. Approximately two-thirds of the Marines remained at PSNS for a time as a Marine Guard Detachment to monitor access to the Controlled Industrial Area or when needed at the outer gates.

Civilian security personnel took over all the Marines' guard duties at PSNS on September 30, 1977, following the final Evening Retreat that evening. All Marine Corps activities in the Pacific Northwest then came under the command of Marine Barracks, Bangor.[6]

The Bangor Annex site was selected in January 1973 to house the Navy's West Coast Underseas Long Range Missile System Support Site. That started the Annex's development to become a major naval installation on the Kitsap Peninsula. Designated in 1974 as the support site for Trident submarines, the area was commissioned in 1977 as Naval Submarine Base, Bangor.[7]

In order to provide space for more quarters at Jackson Park Housing, the buildings and quarters of the former Ammunition Depot on Ostrich Bay were marked for disposal. Some were destroyed, others were sold to individuals and moved to other parts of the county.[8]

Because of their historical significance, a log cabin and stable built by the Civilian Conservation Corps in the 1930s were transferred to the custody of Kitsap County. The county moved them to Silverdale by barge and placed them next to the Kitsap County Historical Society Museum.

A total of 300 units were added to those already available at the Jackson Park Housing. This brought the number of housing units for military families at the site to 700.

A cross-section of a 165-foot tall tree, removed to make room for the new housing, became part of a bicentennial memorial at the intersection of Farragut and Decatur Avenues in the Shipyard. Members of PSNS employee organizations[9] worked as part of the Shipyard's celebration of the United States bicentennial.

An apprentice call in 1973 brought a record breaking number of applicants. Here, during a classroom session, in April 1975, are Electric Shop apprentices (from left) Terry Fagan, Steve McDaniel, Fred Tonnemaker, Cheryl Davies (Stoneking), Mike Ewin, Dale Axtman and Larry Montgomery. Puget Sound Naval Shipyard

The tree section, its growth rings marked to show significant dates in U. S. history, and a chain welded in the shape of the Liberty Bell and bearing the American Revolution Bicentennial Seal face Farragut Avenue. Red, white and blue flowered rhododendrons were planted on the hillside adjacent to Building 850.[10] Dedication of the bicentennial memorial was held on April 29, 1976.

Production Controller John Volpone became the first recipient of an new award established by the Shipyard in June 1975. Although in the 1970s a number of retirees had amassed forty years of Federal service, few had spent 40 years at PSNS. Volpone had; his name was the first engraved on a "plank owner" trophy, signifying he had more Shipyard service time than any other employee.[11]

Volpone entered the Apprentice School in 1935 and graduated as a Loftsman at the rate of $8.48 a day. He became a Planner and Estimator Supervisor before transferring to Combat Systems in 1957.[12]

Jim Reems retired in 1973, after 31 years as editor of SALUTE. Associate Editor Harriet L. McKiernan[13] succeeded Reems. She retired the following year and Philip Judt became editor. In 1980 Judt transferred to the Navy Printing and Publications Service Office and James K. Herron took on the job.[14]

Herron and Electrician Clint Arthur in 1981 organized the Farragut Brass, a band[15] of volunteer musicians, who average over 30 public appearances a year. The Employees Services Committee[16] provides support, enabling the band to provide music in the Shipyard and surrounding towns for occasions as varied as ship arrivals and prayer breakfasts.

Membership was originally restricted to Shipyard employees but soon grew to include those of other commands and military personnel. At one time the roster included four active duty Navy captains, a lieutenant commander, a lieutenant and two petty officers.

PSNS's involvement with nuclear-powered surface ships intensified the need for skilled workers and test engineers; training and recruiting were stepped up. Recently retired employees[17] were recalled early in 1973 and, after qualifying as civil service examiners, traveled throughout the country interviewing candidates and hiring needed personnel.

Often they visited economically depressed areas. On one such occasion the interviewer gave an applicant money to buy lunch while waiting for his interview. Years later the

A Sheetmetal Shop apprentice makes a louvered ventilation duct cover from the drawing at his left. Completed items are complex cross-section ducts and ventilation duct diffusers. Puget Sound Naval Shipyard

cashier at a Bremerton restaurant told the former interviewer that his bill had already been paid, and pointed to the person who had done so. On being questioned, the man said "I've been waiting for a chance to repay you."

The recruiters were empowered to hire; only a security clearance was needed to put the men to work. Later in the 1970s, the interviewing was conducted over "800" telephone numbers, as job seekers answered advertisements.

Former Electric Shop Superintendent, James Pappas, who worked in the program until 1980 explained:[18]

There was a shortage in the Puget Sound area of the skilled workers we needed and pairs of recruiters went as far as the East Coast to interview prospective journeymen and engineers. We hired many excellent workers, who did well in the Yard, but telephoning was much less expensive. Gilbert Burt and I answered phones in the IRO office and if the caller seemed promising we sent an application. That system worked well, too.

Over 200 apprentices representing 23 trades reported at the Shipyard in June 1973 in response to PSNS' largest call for apprentices since the Korean buildup. Included in this number was Gerry Stonecipher, whose brothers Donald, Cleo Jr. and Dale were already apprentice painters.

Donald and Cleo Jr. graduated in 1976, the year their brother Chris entered the program, also as a painter. Their father Cleo Stonecipher Sr. had been in Shop 71 from 1951 until 1963, when war injuries forced his retirement.

The buildup of employees was assisted by the arrival of former employees of the Hunters Point Naval Shipyard which was closed as a national budget reduction measure. The PSNS workforce rose from 9,178 in January 1974 to 10,674 that November.

PSNS offered equal employment opportunities long before being mandated to do so by law in 1972. In 1952 PSNS received its first Award of Merit for employing the physically handicapped.[19] Charlie and Helena Dunkle, both blind, began operation of the Shipmate Snack Bar at the Main Gate in April 1969. As the PSNS' first century draws to a close, almost ten percent of the work force have a disability of some sort.

Shipyard Commander Floyd Schultz established the Shipyard's Equal Employment Opportunity Committee in 1963. Members were: Chairman Loxie Eagans, Shop 72;

An intent Tool Shop apprentice concentrates on the "practical factors" part of his trade. Learning machine tool skills is a major part of his school curriculum. Puget Sound Naval Shipyard

Inserting coil windings in electrical rotating equipment is a precise and painstaking task. This Electric Shop apprentice is preparing to insert a winding in a large stator assembly. Puget Sound Naval Shipyard

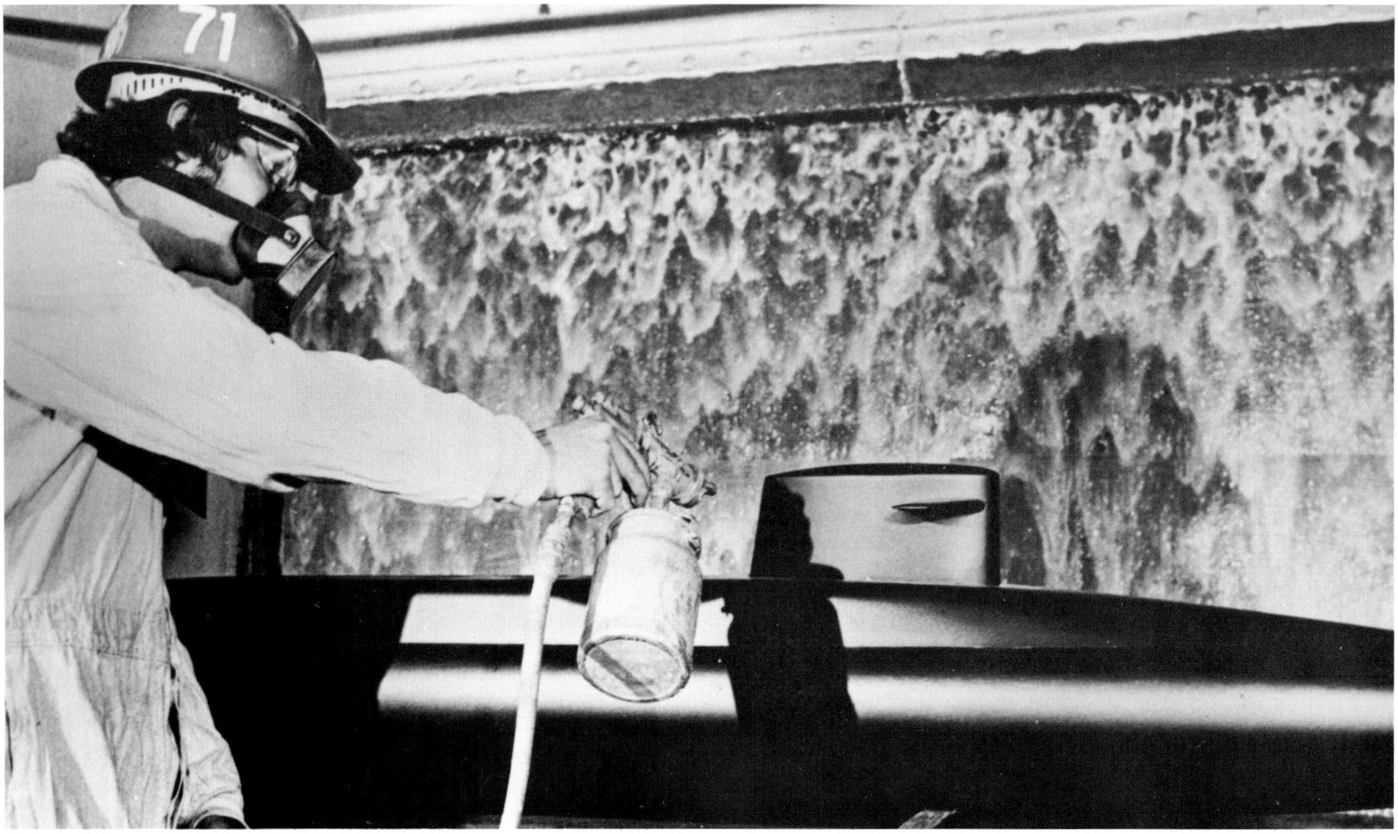

Putting a finish coat on a submarine model, this Paint Shop apprentice is learning careful spray painting techniques. The water curtain in the background captures the paint overspray and carries it down to a floor drain for collection.

Puget Sound Naval Shipyard

Working with older and more skillful mechanics, this Molder apprentice helps with the pour for a stainless steel pump casing. Because of the intense light given off by the molten metal, the three workmen wear very dark protective glasses.

Puget Sound Naval Shipyard

Loxie Eagans, first full-time Deputy EEO Officer at PSNS, welcomes Federal Women's Program Manager Pat Beatty (Caldwell), in 1976 after her return to the Shipyard from a two-year stint at Keyport Torpedo Station.

Puget Sound Naval Shipyard

David T. Lewis, Shop 71; Vicente Barrios, Shop 64; Paul C. Warner, Shop 56 and Ben F. Lyons, Supply Department.

They were responsible for advising the shipyard commander on matters affecting the implementation of EEO policy in the hiring, selection, training and promotion of employees. Eagans was selected as the first full time Deputy EEO Officer in 1969, a position he held until his death in 1981.[20]

Pat Beatty (Caldwell), selected as Women's Program Manager in 1973, was the first full-time women's program manager at a naval activity in the Thirteenth Naval District. As the EEO program expanded, other special emphasis managers and counselors were added.[21]

Women have been employed at PSNS since early in the century. While their numbers temporarily increased dramatically, although temporarily, during the two world wars, the years since 1973 have seen a steady rise in the number of women making careers in both blue collar and professional positions. At the beginning of the 1990s, 15 percent of the work force and almost a quarter of apprentice program entrants were women.

Representatives of most races and nationalities have worked at PSNS, both as civilians and military. In the earliest days, the names of workers reflected mostly northern European origin, but soon Blacks and Filipinos were employees.[22]

Research by Diane Robinson indicates there were more than 200 Blacks in the Sinclair Inlet area by 1912, and most of the men worked in the Navy Yard. Black men have long served in the United States Navy; Chief "Dick" Turpin, who retired in Bremerton, was one. He worked as a rigger/diver in the Yard and his wife Fay was a rivet passer during World War I.[23]

Victor de Gala and Mariano Barrinuevo came from the Philippines in 1910 and worked in the Navy Yard in the Rigger and Woodworking shops. De Gala built a home on the hill above the present Bremerton Sewage Plant site and provided shelter for other arrivals from the Islands. With their help the house grew until it became known as "the castle".

Vicente B. Barrios, long a leader in the Filipino-American Community, started work at the Navy Yard in 1928 as a temporary laborer but took advantage of educational op-

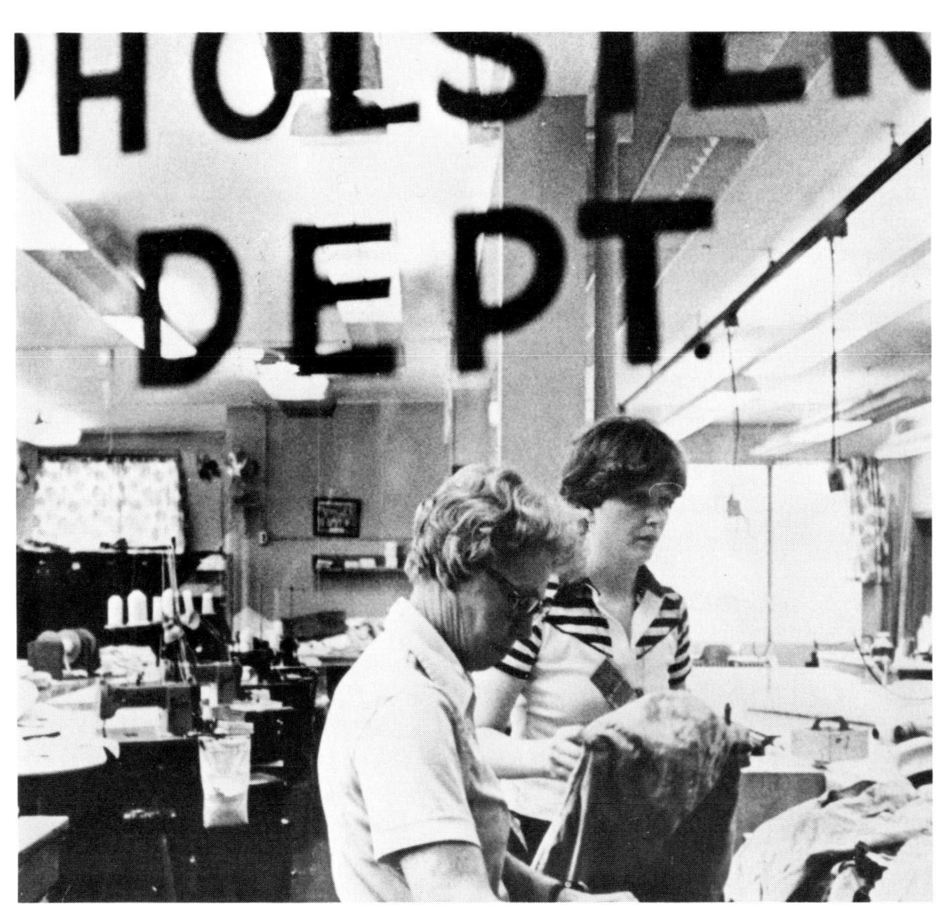

Two photos reflect changes in work conditions for Fabric Workers. Dina Olsen makes gunpowder bags of gray silk in 1925 when the Sail Loft and Loft Riggers shared the second floor of Building 91. During World War II, Sailmakers and Upholsterers moved up to the "clean room" spaces on the third floor vacated by the Director Shop. Later, in 1978, Lavee Reese and Helga Schrub worked in much improved spaces in Building 457, above the Riggers Shop.

Puget Sound Naval Shipyard

Building 50, survivor of the Station's early days, received a new coat of white paint in September 1974, its first since the 1950s. The building located on the southside of Farragut Avenue between Dry Docks 4 and 5 is now used as a ship superintendent's office. Puget Sound Naval Shipyard

portunities in the Yard:

> [Master Shipwright] *MacArthur encouraged me to go to night school in the Yard for two years to become a shipwright . . . I became a shipwright and worked for 34 years before retiring.*[24]

Native Americans working in the Yard during World War II included Wenatchee Indian leader, Chief Kiutus Tecumseh and Daniel Working Bull, the grandson of Sitting Bull. Levi Owens, hereditary Sioux Chief, who worked in the Yard briefly before World War II, returned in 1962 as a welder.

The emphasis of PSNS' work load during this period was on submarines and nuclear-powered surface ships. However, homeported SACRAMENTO and CAMDEN arrived regularly to tumultuous welcomes from families and friends. Referring to SACRAMENTO (AOE 1), CAMDEN (AOE 2) flew a banner declaring "We're number 2 . . . We try harder",[25] when she arrived in February 1974.

The following year, after a month in the Indian Ocean, CAMDEN rushed to the cyclone-stricken island of Mauritius. The crew cleared roads and restored electrical power and communication systems. Ship personnel visited hospitals to donate blood; the ship's cooks baked 500 pounds of bread daily for refugees and orphans.

Submarine tenders HUNLEY (AS 31) and HOLLAND (AS 32) had repair and conversion availabilities in the Yard. March 1973 marked HUNLEY's first appearance in Puget Sound waters. HUNLEY's slogan "Happiness is conversion at PSNS," guaranteed good relations with the Shipyard.

USS BAINBRIDGE was a newcomer to the area when she arrived for modernization in June 1974. At that time she was classed as a nuclear guided missile frigate (DLGN 25). The following year she became a nuclear guided missile cruiser (CGN 25) as part of a major reclassification of ship types. TRUXTUN's designation also was changed to CGN.[26]

PSNS continued to overhaul carriers, but they did not fill the waterfront as they had in the 1960s. CONSTELLATION, KITTY HAWK, RANGER and CORAL SEA all

PSNS' 250-ton hammerhead crane lifts a 230-ton ballistic missile stowage compartment into the empty stuctural space in the hull of USS HUNLEY (AS 31) in May 1973. All feasible mechanical, structural and electrical components were fitted into the new structure which was constructed on a 750-ton capacity barge.

Puget Sound Naval Shipyard

In the ongoing task of automating Supply Department procedures, Code 525.3, Automated Receipt Processing, was opened on July 11, 1980. Shipyard Commander, Captain J. H. Boyd Jr. cuts the ribbon to begin the operation.

Puget Sound Naval Shipyard

USS ENTERPRISE (CVN 65) was in the Yard for a restricted availability from August 1973 to February 1974 and again from January 1979 to February 1982 for a conversion overhaul. She is shown here at Pier 3 near the end of her later availability with steam and utility barges alongside. The string of floats surrounding the ship serves to contain any surface-borne pollutants from the ship.
Puget Sound Naval Shipyard

had availabilities. However the ship that dominated the waterfront at the end of the decade was 18-year-old USS ENTERPRISE (CVN 65), the Navy's first nuclear-powered aircraft carrier. She had a six month restricted availability in Dry Dock 6 in 1973, but returned in January 1979 for an intensive conversion overhaul that continued into 1982.

SALUTE kept its readers informed of progress on the Big E in "the largest overhaul ever attempted in any shipyard and the first complex overhaul of a nuclear aircraft carrier in a naval shipyard."

The overhaul included major engineering improvements, new weapon systems and changes to mast and island. Shipyard forces resurfaced the four-acre flight deck, sandblasting the old deck and applying a non-skid covering. Early in the morning of August 10, 1979, Shipyard cranes lifted the ENTERPRISE's new mast 126 feet before swinging it into place on the ship's 011 level. Captain J. W. Austin placed a Susan B. Anthony silver dollar and other mementos under the foot of the mast in the "stepping of the mast" ceremony.

SALUTE reported fully on "The most comprehensive habitability project ever attempted aboard a Navy ship." General Foreman Aaron Tell's tile setters laid terrazzo and vinyl floors in the Commissary area in early 1980 and the ship's crew began working on interior improvements; almost every living compartment on the ship received attention.

Before ENTERPRISE left the Shipyard, Captain Roger B. Horne Jr.[27] returned to PSNS and relieved Captain John Huntley Boyd as Commander of the Puget Sound Naval Shipyard on June 20, 1981.

The Shipyard's tenth decade was about to begin.

The first view of the Shipyard seen by visitors arriving in Bremerton on the ferry is the east side of the Shipfitter Shop. There, for all to see, is "Puget Sound Naval Shipyard" and the Yard's slogan "Building on a Proud Tradition". Picnic tables and benches in the foreground are part of a small park built in 1987 for employees' lunchtime enjoyment.

Puget Sound Naval Shipyard

Chapter 10

Building on a Proud Tradition

1981-1991

During the 1980s, ships of many classes appeared on the Shipyard's waterfront. The heavy work load consisted largely of submarine overhauls. One submarine, however, arrived early in October 1982 for a short good will visit. This was USS BREMERTON (SSN 698), launched July 22, 1978, at General Dynamics' Electric Boat Division in Groton, Connecticut. The Bremerton High School Band and Drill Team led the welcome for the city's namesake ship on its first visit to Puget Sound.[1]

CONSTELLATION, KITTY HAWK, BAINBRIDGE, TRUXTUN, SACRAMENTO and CAMDEN all returned for overhauls. During a brief visit early in June 1983, SACRAMENTO and CAMDEN shared Pier 4 on one of the rare occasions when both ships were at PSNS at the same time. In 1990 TRUXTUN arrived for her first availability at PSNS as a homeported ship.

Admiral Hyman Rickover made regular visits to the Shipyard in connection with nuclear work. At such times he occasionally briefed the Yard's nuclear power team at a luncheon meeting. He is seen here with Mrs. Rickover and Shipyard Commander, Captain Roger Horne, leaving the Officers' Club. Puget Sound Naval Shipyard

The second ship to bear the name USS BREMERTON is a fast attack submarine with hull number SSN 698. Built by General Dynamics Corporation's Electric Boat Division in Groton, Connecticut, USS BREMERTON came to the Yard on October 15, 1982, for a goodwill visit.

Centennial Book Committee

HIGH POINT (PCH 1), the experimental hydrofoil which had been homeported at PSNS most of its existence, was deactivated in 1982 at the Naval Inactive Ship Maintenance Facility at the Shipyard's west end. Two inactive ships left the Facility when the Navy decided to reactivate the four IOWA class battleships. PSNS prepared USS NEW JERSEY (BB 62) and USS MISSOURI (BB 63) for towing to Long Beach Naval Shipyard for modernization work and recommissioning. NEW JERSEY left in July 1981.[2]

In spite of efforts to retain MISSOURI because of her historical significance, she left in May 1984. The "Mighty Mo" had been a major tourist attraction in Bremerton since 1955. Over the years, an estimated six million visitors stood on the spot where General MacArthur accepted the Japanese surrender to end World War II.

MISSOURI had been the site of events as varied as reenlistments and the filming of Hollywood movies. The ship was closed to the public in August 1976 for filming scenes for the movie "MacArthur" and again in October 1981 for "Winds of War", which was produced for television.

MISSOURI returned to Bremerton in 1988 for a visit. Although the city of Bremerton would like to have her back, her final destination when she completes her active service remains unknown.

A number of code and shop changes occurred in the 1980s. Shipwrights formerly assigned to the Woodworking Shop were transferred to Shop 72 in 1986. Donald M. Sherman remained Superintendent of the enlarged Rigger Shop. The relocation from woodworker to rigger shop was made because a shipwright function is the fabrication and placement of caps on docking blocks; setting blocks is a rigger operation.

Universal Studios came to Bremerton in August 1976 to film scenes for the movie "MacArthur" on the decks of USS MISSOURI, moored at the Shipyard's west end. Many people were eager to volunteer as extras. Local service clubs received the pay of their members who portrayed naval personnel. Gregory Peck, playing General MacArthur, stands at the microphone.

Puget Sound Naval Shipyard

USS MISSOURI (BB 63) is seen here under tow on her way to Long Beach, California, to be modernized and returned to active service. Bremerton-Port Orchard foot ferry CARLISLE is in the foreground making her landing at the Bremerton pier.

Puget Sound Naval Shipyard

Looming over the head of Dry Dock 6, USS CARL VINSON (CVN 70) has the first coat of anti-corrosive paint applied over newly sand-blasted surfaces on her bow. Sand-blasted and repainted chains for her port and starboard anchors are on the drydock floor awaiting re-installation. USS CARL VINSON (CVN 70)

Sandblasting the side of a large aircraft carrier's hull 100 feet above the drydock floor is a formidable task. Working from a hydraulically powered personnel lift platform and completely covered with protective clothing is, however, more efficient and safer than working from a gangway suspended from a crane. Puget Sound Naval Shipyard

Drydocking remains one of the most important functions in any shipyard. Parker Snapp described drydocking in the pre-World War II period thus:

A tiny tug came into the dry dock along with the ship, to nudge it into line. We onlookers were interested to see the Union Jack on the bow lowered as the bow passed over the sill, to show that responsibility had passed from ship captain to shipyard. The skipper of the tiny tug stood with his feet almost against the shoulder of the engineer, who sat just below him, but they communicated by bell signals, just as they would have if separated by three decks and a hundred feet.

The tiny tug had a delightful habit of blowing perfect smoke rings every now and again that rose and hung in the air . . . Side blocks were pulled against the lower bilge of a ship by under-water chains, and shores were propped from the sheer strake to the dock walls on both sides, to make sure that the ship remained erect.[3]

Tugs play an important part in shipyard waterborne operations, moving ships of all sizes in all sorts of sea conditions. Large ships are met on their arrival in Puget Sound by Shipyard tugs under the direction of one of the Shipyard pilots[4] who boards the ship and "takes the con".[5] He directs the ship's movement into the Shipyard, while communicating with the tugs who nudge the ship along the way.

SALUTE was probably quoting a pilot or a skipper of a tug in an article about tugs which had the headline "If it's moving at PSNS, a tug has a line on it."[6] Public Works Department's Shop 02 would undoubtedly question the validity of that remark. Its cranes and other vehicles move throughout the Shipyard's industrial area, transporting and maneuvering material into the needed location. According to Fred Timmerman:

A crane is an inanimate object until it is brought to life by the crane operator. PSNS' were some of the best. The operators of the big cranes were rotated so they could maintain their expertise. One of the first operators I knew was C. J. Sherman . . . He had large strong

Prior to 1935, Building 427 was the Public Works Department's Transportation Building. Here cranes were returned for servicing and refueling at the end of the day's work. Larry Jacobsen

Shipyard photographer Robert D. Balogh, swings above a dry dock and steadies the camera to get a picture for SALUTE. The rigger with him is giving hand signals to the crane operator to hold the skipbox in position. Balogh is now head of the Visual Information Support Center.

Puget Sound Naval Shipyard

hands, which after years of squeezing controls, were like steel . . . Some others I remember were Bushaw, Lambert, Adams, Jack and Fred Haydock, Max Ziegler, Les Holland, Carl Schrieber and Ted Strong. There were others, of course, and they were all good.

Drydocking begins with a "Drydocking order" which specifies dates and times and the vessel's docking plan. The latter details the placement and height of the individual blocks used to support the ship. SALUTE explained drydocking in its February 14, 1969, issue:

Between the opening of the caisson and the dropping of the last hookline a lot of shipyard mind and muscle will have been expended . . . Under the direction of their respective shop docking supervisors who, in turn, work closely with the Shipyard docking officer, the docking force acts as one, in a smooth, coordinated operation that could only be the result of docking savvy and know-how long seasoned in experience.
It is Shop 64 shipwrights and Shop 72 riggers who team up the closest in actual ship docking, with Shop 03 [Public Works Utilities Shop] *personnel backstopping in their handling of all required actions in the pumping, flooding of the dry dock and allied caisson work . . .*

Drydocking operations are generally accomplished at high tide to obtain maximum clearance for the vessel in passing over the sill of the dry dock, and when possible, are scheduled to avoid disruption of other shipyard work. The weather is sometimes stormy and night time operations are particularly uncomfortable and sometimes hazardous.[7]

The next workers to be involved with docking a ship are in Shop 99, Temporary Services. They make the necessary hose and line connections to supply the ship with telephones, power, water, steam, air and drainage facilities while it is in the dry dock.

When a ship is ready to be undocked, the dry dock is partially flooded to allow the ship's company and yard workmen to check for leaks in the ship's hull. The rest of the procedure is similar to docking, but in reverse, with particular care taken at the moment the ship begins to lift from the blocks.

Public Works Shop 07 manufactures the keel blocks used in the Shipyard. They are cast in concrete in cubes that are four feet on each side and are made in large batches. The blocks in use now were made in 1974, the first new keel blocks made since 1940.[8] Any repairs necessary on the dry dock itself are also handled by Shop 07, which is responsible for the upkeep of all structures and facilities throughout the Shipyard.

The only ship in the AOE 1 class of underway replenishment ships not built at PSNS is USS CAMDEN (AOE 2). She is now homeported at Puget Sound Naval Shipyard. Puget Sound Naval Shipyard

In a realignment of shops in the Production Department in 1989, Structural Group, Code 920, was split. Group 920 with Al Burton as Group Superintendent, retained the Shipfitter, Sheetmetal and Welding Shops. New Group 960 included the Boiler, Pipe and Insulator Shops with Jerry Main as Group Superintendent. John Frey, who had been Structural Group Superintendent at the time, was assigned as head of new Code 100Q, Total Quality Management, to spearhead a new concept being adopted nationwide by manufacturing businesses to improve production efficiency.

Shops also experienced changes during the decade. James Davidson became the first Insulator Superintendent. Previously, the work done by the new Shop 57 had been done by the pipe coverers and insulators under the cognizance of the master/superintendent of the Pipe Shop.

Jerry Locke, Superintendent of Shop 41, heads new Shop 42, which was formed to manufacture support equipment for other Production Department shops. Mechanics from various shops are assigned to Shop 42, enabling it to accomplish almost any manufacturing task. The shop accomplishes work formerly contracted out to private industry by the Production Engineering Branch. Production Department's Shop 06, the Central Tool Shop, continues to design, maintain and distribute tools in the Shipyard.[9]

Planning Department's design functions were divided into two divisions in 1987; Code 240 for waterfront design and Code 242 for planning yard design.[10] Robert LaFountaine became Chief Design Engineer for the planning yard division and Leonard M. Anderson for the waterfront work.

A major reorganization of the Administrative Department occurred in June 1987. The Military Support Division and the Budget and Office Equipment Section moved out of the Controlled Industrial Area. Most of the Administrative Services Division was assigned to the Management Engineering and Information Office. The Security division remained in the Controlled Industrial Area, and was retitled the Security Office, Code 1700, with William E. Brady as director.

The unsettled international situation increased the likelihood of terrorism and the need for security. In March the Marines were back on duty in the Yard for the first time in eight years. Nationwide, special badges for entry into Controlled Industrial Areas were required beginning in May 1988, with each naval shipyard having its own badge color. PSNS' is brown. Automated access control systems were installed at all gates except State Street.

Building 91, built in 1903 for the Construction and Repair Department, was declared unsafe in the event of an earthquake. It was razed in 1985.

Building 469, the Heavy Forge Shop since 1942, received

NAVY YARD,
PUGET SOUND, WASH.
MARCH 28, 1914

SHIPFITTERS' AND BOILER SHOP
FOUNDATIONS, CONCRETE PILES
FOR SUPPORT OF WELL COVERING
LOOKING S.W.

WORK DONE
BY YARD FORCE

Perhaps the most graphic demonstration of the Yard's change with time is shown in these two pictures of the Yard Dispensary. This 1914 photo looks over the foundation of Building 178 which was built above the cistern of the large well of the Naval Station. The brick Dispensary (Bldg. 97), in the right center of the picture between Buildings 59 and 108, had previously been used for the Navy Yard's Storekeepers' Office. Kitsap County Historical Society

The present Dispensary, Building 940, modern in both appearance and facilities, was constructed on Farragut Avenue next to Building 850, more centrally located than previous dispensaries. Puget Sound Naval Shipyard

This large hydraulic forging press is one of the main tools of the Forge Shop in Building 452. Handling tongs of various sizes and varied shaped jaws hang from the rack on the pillar at the left. Puget Sound Naval Shipyard

a major remodeling in 1986 and was converted into a propeller repair shop with state-of-the-art equipment. A 2,000-ton heavy forge press and two heating ovens were dismantled. A heat exchange system installed in Forge Shop Building 452 provides hot water to eight shipyard buildings in the vicinity. Heat is recovered from steam previously vented into the atmosphere.

Naval medical services in the Puget Sound area began in 1895 on NIPSIC, moved to Building 51 at the end of the century and to a new frame hospital in 1903. With the completion of the brick hospital in 1911, the industrial dispensary functions of medical care were transferred from the former hospital to Building 97 behind Building 59.[11] From there it moved to Building 56 where it remained until moving into Building 445 in 1938.

A new dispensary (Bldg. 940) opened east of Building 850 on April 2, 1987, to "care for employees who become ill at work; to provide ambulance service; to operate sick call for Navy personnel and to conduct a high level of medical surveillance."[12] The two-story building is the Shipyard's Branch Medical Clinic, housing the Naval Hospital Medical Clinic Staff and the Shipyard Radiation Health Division.

The Shipyard's concern for the health and safety of its employees is reflected in the growth of sections charged with that responsibility. Many people are involved, although the average worker sees only the Safety Inspectors as they patrol the Shipyard on the lookout for hazardous situations or workers without proper safety gear.

Two offices are housed in Building 491, one of the few former hospital buildings still standing. These are the Occupational Safety and Health Office under Shipyard control and the Occupational Health/Preventive Medicine Department of Naval Hospital Bremerton. The latter was formed at the hospital in 1974 under directorship of Commander Richard Nelson.[13] It now has a staff of over 40; its Industrial Hygiene section has grown under Supervisor Roger Beckett to 17 from only 3 in 1963.

The Safety Office was separated from the Industrial Relations Office in 1976 and established as a staff organization reporting directly to the Shipyard Commander. In 1979 this became the Occupational Safety and Health Office, Code 106, under Robert Kieffer. It has a staff of over 40 people compared to only a few in the 1960s.

Hard hats, safety shoes and safety glasses are post World War II developments. When sandblasting replaced hand chipping and wire brushing to clean large areas, personal

The Ground Breaking Ceremony for the Rear Admiral Wallace R. Dowd Jr. Data Processing Center took place on May 15, 1987. When the Center went into operation, its multiple computer complex took over host responsibility for more than 500 customer terminals in the Pacific Northwest and Western Pacific areas. Naval Supply Center

respirators became mandatory equipment where sandblasters were working.

At the Naval Supply Center in the west end of the Yard, a new data processing center (Bldg. 943) was dedicated October 7, 1988. The building was named for a former Chief of the Bureau of Supplies and Accounts, Rear Admiral Wallace R. Dowd Jr.[14] The three story, 41,000 square foot building houses six computer systems that track supply system items.

Sam Josephson[15] entered the PSNS Supply Department in 1962 in the early days of a technological revolution. He wrote in 1990:

In the years that I have been here I have seen the use of computers expanded from that of a payroll processor to a proliferation of desktop models and mainframe terminals which transmit and receive data to and from the four corners of the nation and beyond. Sophisticated computer programs and equipment are used in place of the old punched cards and sorters which used to generate weekly material status and inventory reports.

Now material orders and inventories are monitored and updated around the clock . . . Warehouse issues and receipts are processed in micro-seconds while data of this type in the past was never less than 24 hours old and often much older. The old image of supply folks doing everything manually is gone and operations and record keeping tasks are done with computers and robotics.

New equipment in the Production Department reflects the same trend. Robots are now recognized as essential for dangerous or repetitive tasks in many trades.

This trend is most prevalent in those shops involved in metal working. The Machine Shop began moving in this direction with the acquisition of numerically controlled lathes in 1960. Shop 31 has nine computer controlled lathes which can turn, face, bore, drill, thread, groove and contour a workpiece with repeatable high precision. The latest robotic addition is a 48-inch vertical turning cell which can accomplish machining operations, check the finished piece for tolerance compliance and re-program the work in process.

Welding became robotic with the acquisition in 1990 of a computer controlled welding system that somewhat resembles the general image of a robot. The work is planned

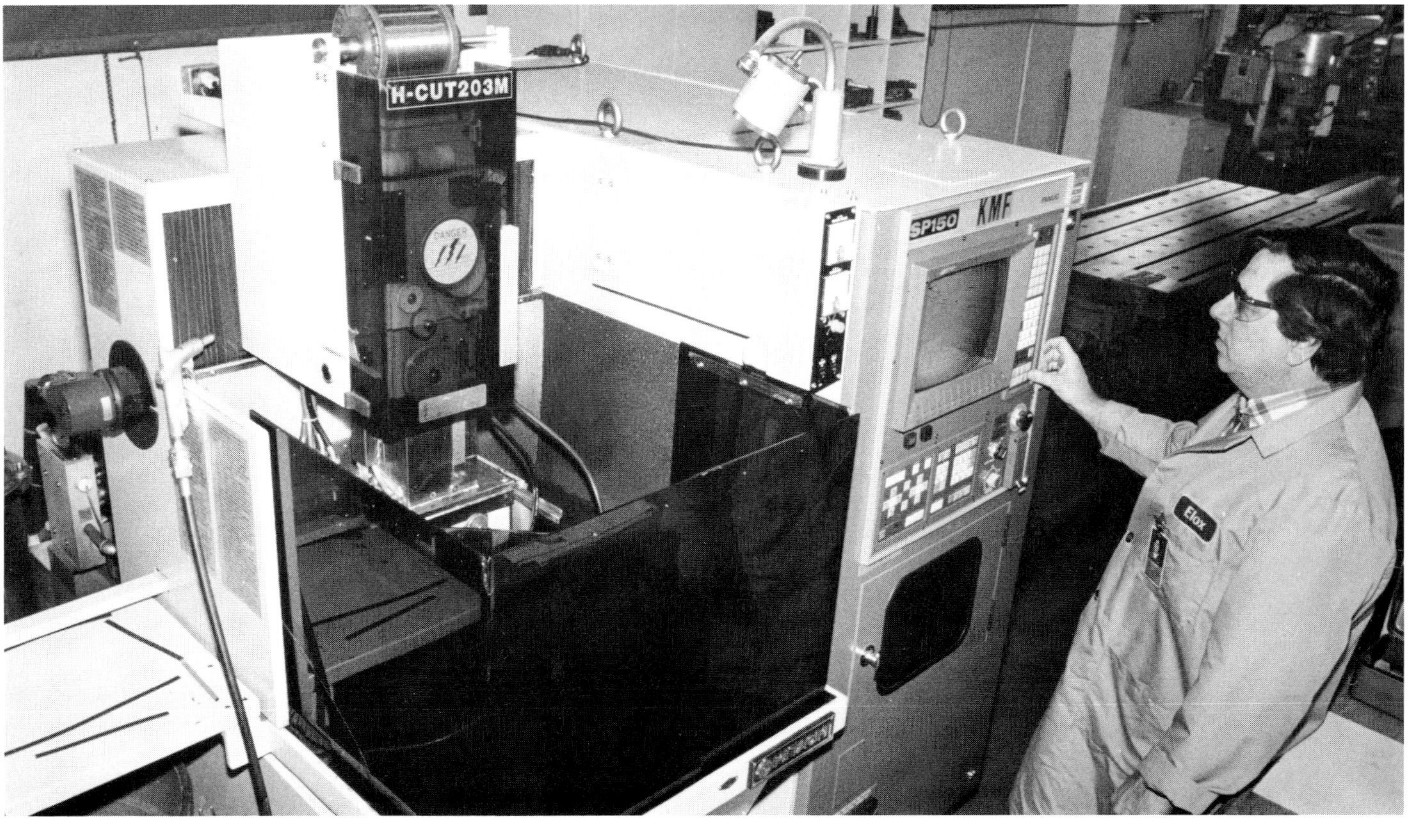

A computer controlled, state of the art, metal cutting machine in the Tool Shop cuts intricate shapes quickly and precisely. The electrically charged wire unwinds from the reel at the top and functions like a tiny band saw as the electrical discharge from the wire cuts a 0.008 inch kerf in the workpiece. The operator is Terry Foxx. Puget Sound Naval Shipyard

The Laser Cutting Center in the Sheetmetal Shop is part of a fully automated CAD/CAM system. The machine is used to manufacture sheetmetal items from stainless steel and aluminum sheets with an accuracy of 0.004 inch. Don Williams is the mechanic inside the curtain and supervisor Rudy Ohlund is overseeing the work. Puget Sound Naval Shipyard

Programmed moves are checked on the robotic welder after being set up to weld a steel container. The welding fixture is seen on the table below the welding torch. Gary Olsen is operating the remote control for the welder. Puget Sound Naval Shipyard

New methods to protect ships' propulsion shafts from sea water corrosion are used in Shop 31. The shaft is wrapped with fiber glass cloth and coated with an epoxy resin for a watertight seal. At work are: clockwise from left, Sid Litman, Carmen Lapid, Frank Burton, Cecil Heiskell and Robert Snyder.

Puget Sound Naval Shipyard

The newest addition to the Machine Shop's automated tools is a Giddings and Lewis 48-inch vertical turning cell. Because it is controlled as a unit its multiple work station must be aligned to a common base line. Here, Tool Shop maintenenace mechanic Ed McClellan is aligning the system optically. Puget Sound Naval Shipyard

The advanced computer control system for the 48-inch vertical turning cell, shown here with the major assemblies in place during the system testing phase. The electrical wireways between the units gives an indication of the complexity of information which passes from the operating consoles to the tools. Puget Sound Naval Shipyard

All of the 1912 Central Power Plant's functions have now been taken over by the new steam plant. This architecturally handsome building, shown during construction in 1909, is now scheduled for demolition. Puget Sound Naval Shipyard

and programmed by computer and accomplished automatically by a single operator. Building 857 now has a laser cutting center, which is integrated with a CAD/CAM system. Sheets of metal up to 1/4 inch in thickness can be cut with this system.

The Tool Shop has a computer controlled electrical discharge wire cutting machine which can be programmed to cut intricate metal shapes and angles with tolerances of a few thousandths of an inch. To maximize safety, the work area is enclosed while the machine is in operation.

The largest recent addition to the Yard is the new Steam Plant in the southwest end of the Shipyard, its 304-foot stack towering above the shipyard and Bremerton. Captain Arthur Clark, the Shipyard Commander, accepted the new steam plant from Rear Admiral Benjamin Montoya, Commander, Naval Facilities Engineering Command, on August 28, 1989.

Plans for construction of the plant were initiated during the oil crisis in the late 1970s; work began on this extremely complex coal-fired plant in 1982. Coal to stoke the fires comes to the plant in 100-ton bottom dump railroad cars; a conveyor system moves it to storage or to the 135-ton capacity bunkers for immediate use. Diesel oil can be used as a backup fuel and RDF, Refuse Derived Fuel, may be burned in the future. RDF is a dry pelletized form of waste, processed into fuel, which burns efficiently and without odor.[16]

Three 140,000-pound per hour boilers produce steam which is distributed throughout the Shipyard for industrial and shipboard use and for climate control in buildings. Hot flue gases from the furnaces are purified in several stages before being released to the atmosphere through the corrosion-resistant stainless steel-lined stack.

The plant and associated distribution systems are operated and maintained by Public Works Department's Shop 03, which is responsible for shipyard utilities such as distribution of electrical power, steam, compressed air and salt and fresh water.

Shop 03 also maintains and operates dry dock pumps and pumpwells. These subterranean pumping stations are capable of dewatering a dry dock at rates exceeding 456,000 gallons per minute. Building 52, whose boilers had provided power for pumping Dry Dock 1, has been demolished. Building 923, built in its place, houses five compressors which supply compressed air in the east end of the Shipyard.

Building 106, the 1912 Power Plant, is scheduled for demolition. In addition to providing steam, the "Power

Pictured here is an artist's rendering of the new steam plant at the west end of the Yard. The large building complex, left center houses the steam generators, machinery and central control room. In the foreground a string of coal cars is shown entering the coal dumping shed from which the coal is transported into the coal handling system. Puget Sound Naval Shipyard

House" had supplied all the Yard's electrical needs until new dams and other power sources made it cheaper to buy than generate electricity. Its generators, however, were kept operational for emergency use. The new plant's five diesel generators have now taken over and can provide 11,500 kw of emergency power.

Since World War I, the west end of the Shipyard has been used primarily for military support purposes. During two world wars it was home to sailors entering or leaving the Navy.

Now, three large high rise bachelor quarters provide living accommodations for enlisted personnel assigned to the Shipyard, tenant commands and ships undergoing overhaul or repair. The first, Building 865, was built on the north side of Cole Street[17] downhill from the still-visible fifth tee of the Yard's second golf course.[18] These quarters were formally named Underwood Hall on Memorial Day 1978, honoring Bainbridge Island hero, Perry Luke Underwood, who was killed in action in Vietnam.

Underwood Hall provides Chief Petty Officers' quarters on floors six through ten. Female quarters are on the fifth floor and temporary family quarters on the first floor.

The first of two Bachelor Enlisted Quarters erected on the hill vacated by the naval hospital in 1981 was dedicated on Pearl Harbor Day, December 7, 1984. Keppler Hall honors Washington-born Reinhardt John Keppler who was posthumously awarded the Medal of Honor for heroism while serving on USS SAN FRANCISCO during the Naval Battle of Guadalcanal.[19]

The second, Nibbe Hall, was dedicated September 24, 1986, in memory of Civil War Medal of Honor recipient John H. Nibbe. Washington's Secretary of State Ralph Munro, whose family settled on Bainbridge Island near the Nibbe homestead, was the principal speaker. Pearl Harbor Medal of Honor recipient, Captain Donald Ross, led the singing of "God Bless America".

Rear Admiral Roger Horne Jr. on February 16, 1988, presented the shipyard with the Admiral Elmo Zumwalt Award in recognition of its having the best bachelor enlisted quarters complex in the Navy. This marked the first time that a shipyard had won the honor. Horne praised the "Every patron a VIP" philosophy of the bachelor enlisted quarters' staff. The Shipyard was first runner-up in 1989 and won again in 1990.

Two new bachelor enlisted quarters, Keppler Hall and Nibbe Hall stand on the site formerly occupied by the Naval Hospital. The staff's philosophy of "Every patron a VIP" continues to make these and Underwood Hall the best enlisted quarters in the Navy.

Puget Sound Naval Shipyard

Ralph Munro, Washington Secretary of State, stands beside his father, George, with his uncle Gordon and a portrait of John Nibbe. Ralph Munro was the principal speaker at the dedication of Nibbe Hall September 24, 1986.

Centennial Book Committee

Late model cars line up at the grand opening of the new gas station in the northwest corner of the Shipyard in August 1972. The new location was able to accommodate the lines of waiting patrons without causing congestion.

Puget Sound Naval Shipyard

Construction of the enlisted quarters marked the first of many major changes in the west end of the Shipyard. The Great Framemaker picture frame shop, Big Al's Pizza, Deli and Grill, SATO commercial travel agency, Legends Sports Bar and Northwest Outfitters followed.

Some war-time structures remain. The World War II Naval Barracks Subsistence Building (Bldg. 464) became a commissary in 1963. Part of the Recreation Building (Bldg. 502), erected during the same war, served as a Navy Exchange until replaced by a new one in 1974. This facility received the Bingham Award in 1989 for the best Exchange of its size in the Navy.

Since the days of Commandant Burwell, much of the area has been used for recreation. It now has playing fields, a swimming pool, bowling alleys, a library, hobby shops and clubs. The former Marine Enlisted Men's club in Building 434 was converted into the Legends Sports Bar in the fall of 1990. A Chief Petty Officers' Club operated for many years near the Yard's western fence until a new one was erected east of Underwood Hall.

In 1990 the Shipyard won the Golden Anchor Award for the quality of its Enlisted Retention Program and the Bronze Hammer Award for Self Help, which recognized the contribution of assigned sailors in the development and execution of various improvement projects throughout the support area.

The growth of the recreational program is assisted financially by a recycling program operated by sailors attached to the First Lieutenant's Division. Collection of aluminum cans, cardboard, paper and other recyclables raises revenue which goes directly into recreational programs and facilities. The program also substantially reduces trash disposal costs.

Administrative offices for military matters are housed in Building 853, across Cole Street from Underwood Hall. The building houses a large Childcare Center and the Family Service Center which is staffed by certified civilian counselors. A full range of support services is available to active duty and retired military and their dependents.

Members of the Navy Morale, Welfare and Recreation Office arrange excursions, provide rental equipment for sports, camping, and recreation for civilian shipyard employees as well as for military personnel.

The growth of support services matched the rapid increase in numbers of military personnel at the Shipyard resulting from homeporting of three more ships, USS TRUXTUN (CGN 35) and two large carriers, USS NIMITZ (CVN 68) and USS CARL VINSON (CVN 70).[20] The arrival of VINSON in 1990 brought the enlisted population of the Shipyard to more than 10,000.

In the words of Commander John Gordon, the Administrative Officer:

The rapid growth of the Shipyard and decline of downtown Bremerton and associated recreational facilities provided both the impetus and necessity for improvement and increases in recreational and support services. Indeed, the Shipyard finds itself providing all the support services normally associated with a major naval base.

The Marine's stable, built in 1913, stood alongside the fence just west of the swimming pool. Before World War II it was the Warrant Officers' Club, later becoming the Chief Petty Officers' Club. It was was used in that capacity until the new Chief's Club opened in the mid-1970s.

Puget Sound Naval Shipyard

The new Chiefs' Club on Cole Street is located in the Shipyard's northwest section, an area covered by the Navy Yard golf course before World War II.

Puget Sound Naval Shipyard

Moments after Bremerton Armed Forces Queen Jennifer Frey christened the refurbished and renamed Bremer-Wyckoff Building she receives applause from other participants in the ceremony. The dedication plaque now stands in the building's lobby.

Puget Sound Naval Shipyard

Space in the Yard became increasingly scarce. Empty business buildings in the city of Bremerton made it possible for the Shipyard to expand outside its fence. PERA (CV) moved into the Manette School on Bremerton's east side in 1984. PERA is the acronym for Planning and Engineering for Repairs and Alterations. The CV indicates the work is done mainly for aircraft carriers.[21]

PERA began at the Shipyard in February 1967 with official word from NAVSHIPS[22] assigning PSNS the task of conducting a study in depth of applicability of PERA concept to CVA overhauls. Design Code 243 was set up with Bruce Towne as Program Manager. It became a separate office in the Shipyard organization in July 1970, and now, as part of the Naval Sea System Command, is responsible for planning on all carrier overhauls in Naval Shipyards.

Engineers and technicians from two codes of the Design Division's Planning Yard section moved into the former Bremer Department Store building on Pacific Avenue in Bremerton in April 1988. The remodeled structure was named the Bremer-Wyckoff Building. Plans were made for the rest of the Planning Yard section to move into downtown Bremerton buildings at a later date.[23]

Pins designating years of federal service are still presented to employees. Many who receive 40 year pins have earned some of that time in the military or at other federal agencies. Among those employees who served over 40 years at the Shipyard before retiring in the past ten years are Harold Pickard, Albert G. Novone, William Boldrin and Wayne Matz.[24]

Two women have worked at PSNS over 46 years: Virginia Thorpe[25] already had five years of federal service in South Dakota when she entered the Planning Department on April 16, 1943. Pearline Giggans[26] began her career in the Personnel Department September 17, 1943.

Lieutenant A. B. Wyckoff, founder of the Shipyard, was honored in 1988 with his inclusion in the Washington State Historical Society's Hall of Honor, a list of 100 people who made major contributions to Washington State history. This was in connection with the celebration of the Washington State Centennial in 1989.[27]

In connection with that celebration, the Bremerton Main Street Association and the Bremerton Parks Department spearheaded a drive to refurbish the Ambrose B. Wyckoff Park at the south end of Pacific Avenue. A large crowd, including Shipyard Commander Arthur Clark and Bremerton Mayor Louis Mentor, attended the rededication of the enlarged memorial on May 18, 1990.

In the group were three descendents of the founder of the Shipyard, including Mary Muir, who as Mary Wyckoff Rogers had attended the original dedication by the Kitsap County Historical Association in 1951.[28] Also attending was Isabelle Birkenfeld, daughter of Harry H. Keith, who prepared the pattern for the memorial plaque when the park was first dedicated.[29]

As the Shipyard's centennial year approached, celebration plans increased. Louise Ashenberg, granddaughter of

Lieutenant Wyckoff's great-granddaughters, Liz Rischell and Mary Muir stand, with George Rischel, to left of Wyckoff Memorial Plaque at rededication of Wyckoff Park on May 18, 1990. Great-grandson Ted Rogers is at right with his wife, Jane. Louise M. Reh

Bremerton Mayor Louis Mentor and Isabelle Birkenfeld stand beside the monument in Wyckoff Park. Mrs. Birkenfeld's father, Shipyard Patternmaker H. H. Keith, made the pattern for the bronze plaque in 1951. Richard Linkletter

Mary Wyckoff Muir, great granddaughter of Lieutenant Wyckoff, prepares to address crowd at the dedication of newly enlarged Wyckoff Park on May 18, 1990. Centennial Book Committee

Assisted by Yard tugs and surrounded by a welcoming fleet of small boats, USS NIMITZ returns home. Visible over NIMITZ' flight deck are the east end of the Yard, the ferry terminal, downtown Bremerton and the Manette bridge. This is a very different panorama than was seen when NIPSIC arrived in 1892. Steve Zugschwerdt, SUN

E. A. Wall, the Navy Yard's first coppersmith, designed the centennial logo. Buttons and other items bearing the logo were sold to raise money for the celebration.

The Puget Sound Naval Shipyard Centennial Committee, chaired first by John Frey and later by Bob Bray, decided on a Bremerton waterfront memorial as a lasting legacy of the centennial year. They chose Bonney Lake sculptor Larry Anderson's design and commissioned him to have his model cast in bronze and installed on the new Bremerton Waterfront Plaza. Money for the project was raised through the sale of bricks and granite squares to be used for paving and decorating the site.

Chapter 14 of the Federal Managers' Association[30] voted in 1990 to act as publisher of this history of the Shipyard. FMA and other civilian professional organizations, such as the Superintendents' Association, National Civilian Managers' Association and National Association of Naval Technical Supervisors support the Shipyard through their programs.[31]

As the Shipyard's first century draws to a close, PSNS is home to many ships, including the second ship referred to in this book's title, USS NIMITZ (CVN 68). When she arrived at PSNS July 2, 1987, NIMITZ was welcomed as enthusiastically as USS NIPSIC had been in 1892. This time the crew's families joined the welcoming crowd. The ship's homeport had been changed from Norfolk, Virginia, to Bremerton, until such time as a proposed homeport base at Everett, Washington, would be completed.

At another arrival in February 1991, a dark cloud dimmed NIMITZ reception. The country was at war; all realized the ship probably would be sent to the war zone. However, when NIMITZ left for the Persian Gulf a few weeks later, hopes were high the ship would return in time for the Shipyard's centennial celebration on September 16.[32]

At the close of its first century, the Puget Sound Naval Shipyard is the Navy's foremost shipyard.[33] It continues "Building on a Proud Tradition".

Lieutenant Wyckoff's dream has come true.

Outside Shipfitters demonstrate removal of a dent in ship's side with the use of a bolting-up jack in May 1920. Man to the right of the jack is Master Outside Shipfitter George Gregoire who began work at the Navy Yard in 1910 and became Shop Master in 1919. When the 41-year old Master died in August 1923, his shop was closed for the funeral and the Navy Yard Commandant, Rear Admiral J. A. Hoogewerff, gave the eulogy. *Puget Sound Naval Shipyard*

NOTES

CHAPTER ONE

1. NIPSIC arrived in Puget Sound in March, but transfer of command was not made until April.

2. Quartermaster John Nibbe, Civil War Congressional Medal of Honor recipient, fired an old Russian cannon from Sidney (now Port Orchard). Nibbe, the area's first entrepreneur, operated stores at Crystal Springs, Mitchell Point and Bremerton. He owned the ships which carried the wares and also served as Sidney's postmaster.

3. Lieutenant A. B. Wyckoff, U.S.N., Reminiscences of the Survey of Puget Sound and the Establishment of the Puget Sound Naval Station. *THE WASHINGTON HISTORIAN,* January 1901, page 61.

4. Ibid., page 57.

5. Ibid., page 58.

6. Ibid., page 61.

7. *THE KITSAP PIONEER,* April 9, 1891.

8. Unpublished treatise by "Class of 1947-1948", believed to have been compiled by the Yard's apprentice class of that date.

9. Unpublished memoir of Mrs. Lewis Bender and undated *BREMERTON NEWS* clipping, quoting Thomas Ross, early Kitsap County Auditor. Dogfish Bay is now known as Liberty Bay.

An 1897 view of the Naval Station shoreline shows Building 50, the Station office building and boats from USS OREGON beached in the foreground. Continuing to the right are the dispensary, (Bldg. 51), the dry dock and the power plant, (Bldg. 52). Under construction are Buildings 58 and 59. Kitsap County Historical Society

10. *THE WASHINGTON HISTORIAN,* January 1901, page 62.

11. Holbrook started work at the Station in December 1892 and from 1899 until 1903 was in charge of all civil engineering work at the Navy Yard. When a Civil Engineer was again assigned to the position, Holbrook continued as a draftsman until his retirement in 1916 at age 76. He left behind a legacy of official weather and tide reports and scrapbooks filled with clippings of newspaper and magazine articles regarding the area.

12. Daniel Lund served under Wyckoff in the Navy. On the expiration of his enlistment he was hired as the Station's first civilian laborer.

13. George Terrill accompanied Wyckoff from Illinois to Kitsap County in 1891 and worked in the Yard until 1910. He was active in real estate and municipal affairs in Charleston (West Bremerton).

14. *THE WASHINGTON HISTORIAN,* January 1901, page 63.

15. Grulich Architecture and Planning Services, Tacoma, Washington, HISTORIC SURVEY PUGET SOUND NAVAL SHIPYARD, Bremerton, Washington, October 1985.

16. An unidentified Seattle newspaper article in album belonging to Lieutenant Wyckoff's youngest daughter, Alice. Article describes Dry Dock 1 and states, "These dimensions probably entitle it to rank third in size among the single docks of the world, and it is certainly the largest in America."

17. Frederick Forbes and his wife built their home inside the Navy Yard fence in 1903. When he retired in the mid 1930s, the Navy Yard remodeled the house and it is now an officer's quarters. Forbes was an excellent golfer and represented the Navy Yard at golf tournaments throughout the Northwest.

18. The other four original quarters are A, B, D and E, which were built for $4,000 each. Major changes through the years have been the converting of each home's two side-by-side parlors into a large living room, and the addition of bathrooms.

19. *THE BREMERTON NEWS,* August 15, 1916.

20. *BREMERTON SEARCHLIGHT,* Special Navy Yard Edition, undated.

21. *THE WASHINGTON HISTORIAN,* January 1901, page 63.

CHAPTER TWO

1. The other plat was Charleston; the two combined as Port Orchard, the collective name given the bodies of water west of Bainbridge Island, between Agate Pass and Rich Passage. However, the small town of Sidney on the south shore of Sinclair Inlet had already petitioned for a Post Office of that name. Until 1903 the town of Sidney had the Port Orchard Post Office and the town of Port Orchard had the Charleston Post Office. An act of the State Legislature changed the names of the two towns to match their post offices.

2. There had also been confusion about the name of the Station. Many referred to it as the Port Orchard Naval Station, hence the desire to have that name for the towns.

3. Ray Raines was the son of Thomas Raines, who brought a group of painters from Mare Island to work on USS MONTEREY at the new Puget Sound Naval Station in 1896. When he became "Quartermaster Painter in Charge" at the Navy Yard in 1902, he moved his family to Bremerton. The family lived on what is now the 1600 block of Burwell Street until 1916 when they returned to California, where Thomas Raines became Master Painter at the Mare Island Navy Yard.

4. The Steam Engineering Department was in charge of all machinery on the ships. Jesse "Red" Jones was born aboard NIPSIC on April 12, 1900. His father served at the Navy Yard during World War I and "Red" worked there later as a machine operator.

5. *MANETTE PIONEERING*, page 278.

6. Thomas Bright's son, Arthur, worked in the Yard's Sheetmetal Shop for 38 years, retiring as a Chief Quarterman. His son, Lennox, worked in the Machine Shop.

7. Construction and Repair Department was in charge of all hull work.

8. The Navy Yard had electricity before any of the surrounding towns.

9. Quarters "J" was the first of four Warrant Officers' quarters erected near the beach. They were later moved up the hill and became Commissioned Officers' quarters.

10. Douglas was Foreman of the Riggers and Laborers. When he died suddenly in 1906, *THE BREMERTON NEWS* described him as ". . . built upon a broad and liberal plan, ample without limitations, gentlemanly not servile, courteous yet dignified — he ever had the courage to perform his duty as it appeared to him."

11. *BREMERTON SEARCHLIGHT*, August 5, 1905.

12. Ibid.

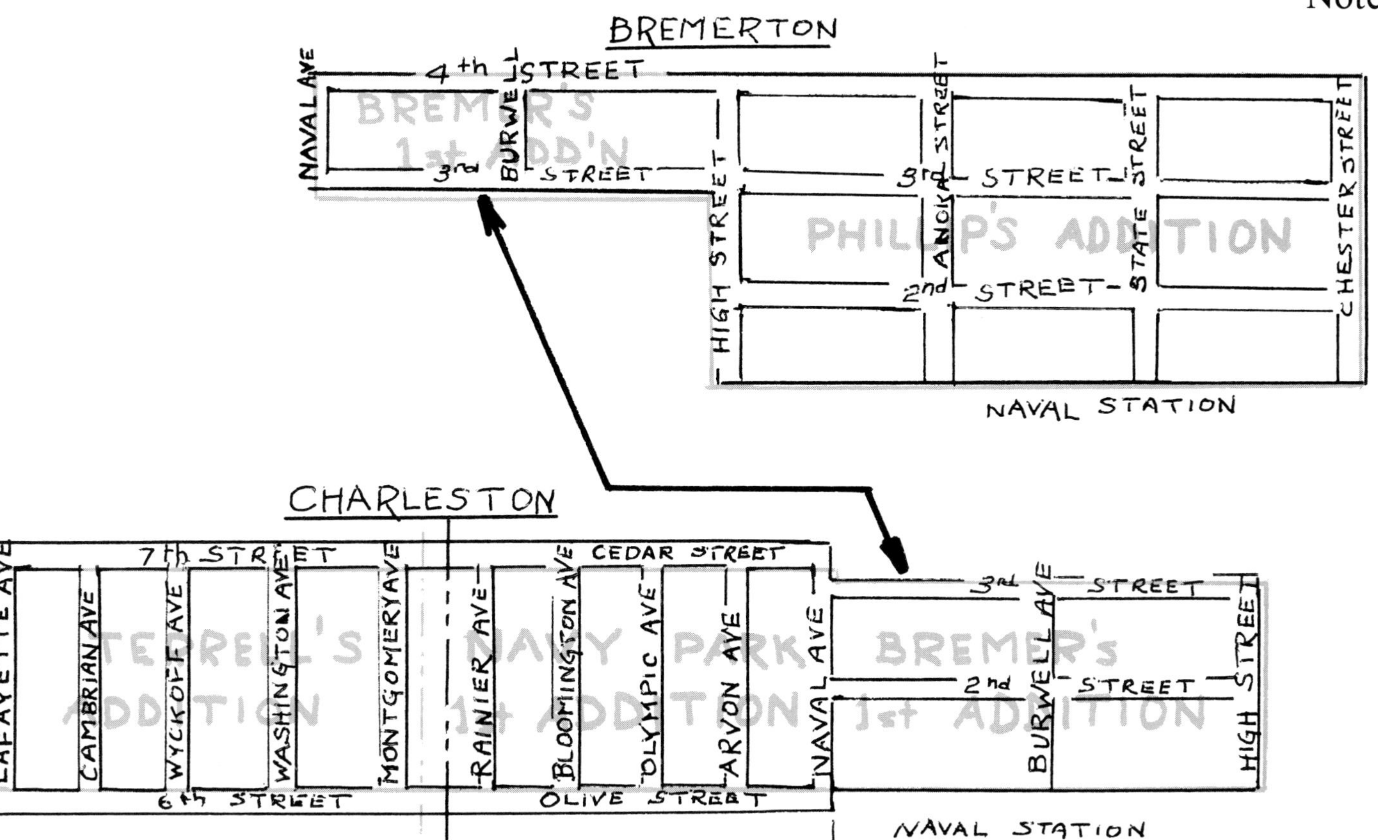

Early plats reveal how Burwell Street was developed from streets in Bremerton and Charleston. Plats and their dates of filing are Terrell's Addition to Charleston, October 3, 1892, Navy Park First Addition to Charleston, June 14, 1891, William Bremer's First Addition to the cities of Bremerton and Charleston, January 3, 1906, and Phillips' Addition to the Town of Bremerton, June 28, 1901. Wyckoff Street is spelled correctly in The Terrell Addition, but was incorrectly spelled on later plats.

Michael Mjelde

13. *SEATTLE POST-INTELLIGENCER*, January 14, 1912, article and Wyckoff letter in University of Washington Manuscript file, addressed to Speaker of the House Joseph Cannon, indicates Wyckoff also assisted in getting approval for Dry Dock 2. In both, Wyckoff stated, "Speaker Cannon said to me 'God Almighty made that location for a Navy Yard', and at the next session of Congress he repeated this remark in a speech on the floor of the House in favor of the appropriation for another dry dock, and it was passed."

14. Sophia and William Bremer sold .40 acre to the government for $500 in 1896 and 1.53 acres for $1,000 in 1900.

15. The pictures on pages 26 and 27 are often mistakenly labeled "The Great White Fleet". The pictured ships are the armored cruisers of the Pacific Fleet which was in the Navy Yard the month before the Atlantic Fleet. Following the completion of the "Great White Fleet" voyage, Navy ships were painted gray. People, remembering the whiteness of all the ships, gave all of them the name now limited to the globe-circling Atlantic Fleet.

16. USS NEBRASKA, built by Moran Brothers of Seattle in 1904 and outfitted and commissioned at the Puget Sound Navy Yard, joined the fleet to replace USS ALABAMA.

Page 27. Last paragraph should read ". . . gave Burwell's name to Third Street in Bremerton and Cedar and Seventh Streets in Charleston."

17. Part of this land was transferred to Bremerton in 1967.

18. *BREMERTON SUN*, November 14, 1951. One boy who played with the band was Harry Jertson, who was born in the family home shortly before Lieutenant Wyckoff completed purchase of the acreage from the Jertsons.

19. Truman H. Newberry, 39th Secretary of the Navy.

20. Land on Ostrich Bay was purchased in 1904 for a naval magazine.

21. The original Wireless Station, Building 133, is now an officer's quarters.

22. *SEATTLE POST-INTELLIGENCER*, January 5, 1910. Another death was that of William Bremer in December.

23. Cottman was well known in the Navy Yard. He had been Commanding Officer of both WYOMING and CALIFORNIA and served as Captain of the Yard (second in command) for Barclay, Burwell and Rodgers.

24. BRIEF HISTORY OF THIRTEENTH NAVAL DISTRICT states, "On 26 July 1910, the joint" (12th and 13th) "district was divided establishing the Thirteenth Naval District." Register of Officers for 1909 and 1910 confirms this.

Page 32. Caption on photo should read, "Its great weight of **granite**".

25. Earlier histories give the depth as "38 feet deep at high tide." This refers to the depth of water over the dry dock sill at high tide. The actual depth of the dock from ground level surface to dry dock floor is 46 feet, according to the Grulich report.

26. The Veterans' Home was named Retsil, Governor Lister's name spelled backwards.

Page 33. Caption on photo should read "on March **1**, 1913."

27. Chief of the Bureau of Yards and Docks R. C. Hollyday letter regarding transfer of land: *In this connection the Bureau will state that in 1896 while the present Chief of the Bureau was Civil Engineer at the Puget Sound Navy Yard* [sic] *the question of a reservation for the Yard* [sic] *came up and at the time he strongly recommended that the land in question be turned over to the Marine Corps for drill ground, etc. but the Inspector of the Marine Corps at that time insisted upon what is now the Marine Reservation being set aside for the Marine Corps.*

28. Marine Corps Commandant Joseph H. Pendleton, for whom Camp Pendleton in California is named, was commanding officer of the Marine Barracks at PSNY from November 1906 to September 1909 and September 1913 to December 1914.

Page 37. Paragraph beginning "PSNY received a 600-ton anchor chain" means the 600 tons refers to the hydraulic testing force the machine can apply to the anchor chain links.

29. Building 157 was constructed for $260,000. Source article states that Contractor C. F. Graff would also build Pier 8, "a steel concrete structure which has never been undertaken on this coast before."

30. The Torpedo Station is now the Naval Undersea Warfare Engineering Station.

31. Coontz was a strong supporter of the towns around the Navy Yard. A street in Charleston was named for him and when Bremerton High School was built in the late 1940s, Union High School became Coontz Junior High School.

CHAPTER THREE

1. One of the advantages stressed in the early days of rivalry between PSNY and Mare Island Navy Yard was the fact that so little dredging is needed in Sinclair Inlet.

2. Charles Eugene Talmadge, THE GROWTH OF PUGET SOUND NAVAL SHIPYARD, page 45.

3. Captain Gregory, Public Works Officer in the Navy Yard from July 1913 to May 1920, was honored when Bremerton renamed Second Street Gregory Way.

4. Following her father's death in the mid 1930s, Bertha Coontz worked in the Navy Yard.

5. Secretary of the Navy Report, 1921, page 91.

6. Alice Bender Davis letter to Louise Reh dated November 29, 1985. Union High School was built in 1907/1908 between Fourth and Fifth Streets, on property half in Bremerton and half in Charleston.

7. More than 200 Yeomanettes can be counted in photo taken in front of Quarters C in 1919.

Page 45. Cutline on photograph should read "Standing to his right is Lyle **McLeod** . . ."

8. Robert E. Coontz, FROM THE MISSISSIPPI TO THE SEA, page 385.

9. SECRETARY OF THE NAVY REPORT, 1918, page 1489.

10. Gavin is author of unpublished (as of 4/91) AMERICAN WOMEN IN THE GREAT WAR. Burns was Mrs. McCrary when she was a Yeoman (F). After the war, she was Deputy Clerk of Jefferson County for 30 years and later Curator of the Historical Museum at Port Townsend.

11. Member of pioneer McGowan family, enlisted as a Yeoman (F) when she was 18 years old and was secretary to the Chief Draftsman. She married Joseph Madden, whom she met during the war.

12. Female Nurse corps was established in 1908

13. SECRETARY OF THE NAVY REPORT, 1919, page 2291.

14. Julia Hippe, quote in *GLENWOOD HERALD*, Glenwood Minnesota, November 29, 1917.

15. Gavin quotes Fitzgerald, "It was such an honor to be one of the first women in the Navy, it was a beautiful part of my life."

This November 13, 1916 photo shows Burwell Street in the foreground and Park Avenue to the right, looking into what had been Burwell Park. When Bremerton opened Evergreen Park, the Burwell Park land was returned to the Navy; tennis courts replaced the band stand and children's playground. The Telephone Company Building, now the Chamber of Commerce, northwest of the courts is still standing at the location facing Fourth Street. Puget Sound Naval Shipyard

16. *SMITHSONIAN*, January 1989, pages 130-146.

17. See text on page 80.

Page 48. Last paragraph in left column should have indicated the hour long trip from Seattle was by ferry. The trip by land would have taken longer.

18. SECRETARY OF THE NAVY REPORT, 1918, page 655.

19. Sand Point Archives, Entry 69-70691, Admiral Coontz' papers. Blueprint shows purchases as follows: Bremer 10 acres, Jertson 6.5 acres, McDonald 1 acre, Moe 0.85 acre, Sinclair 0.5 acre and King 0.5 acre. This land was developed into the third through seventh holes of the Yard's second golf course, until barracks buildings were constructed there during World War II.

20. Although District-related work was done in Seattle after 1917, the Commandant continued to live in Quarters C and have an office at the Navy Yard.

21. Contrary to popular opinion, the hull that was buried at Evergreen Park for many years was not of a submarine built at PSNY, but of USS FOX, a torpedo boat built in Portland, Oregon in 1898. It was only a few feet different in length and inches in beam from the submarines, but its flat deck and light hull plating identified it as a surface ship.

22. Information received from Captain Q.E.D Lewis(CEC) USN (Ret).

Page 53. Caption of photo at bottom of page should read: ". . . as well as two target rafts which . . ."

23. Mrs. Gregory, the former Pauline Turner, was an accomplished vocalist. Her brother was a professional photographer whose name appears on many photos of the Navy Yard.

24. Linkletter biography of Winsor: "Entered PSNY as draftsman in 1908, became supervisor of planning and estimating section he helped organize. During World War I, Winsor was head of progress section for scheduling and expediting work and he established the quality assurance laboratory as a test lab in Electric Shop in 1919."

25. Linkletter biography of Cowles: "Formerly private contractor and field engineer for Westinghouse Electric Company; came to PSNY in World War I to help war effort and set up test lab with Winsor. Electric Shop supervisors under his direction were Maurice Allen, Walter McConnel, Ralph Seigner, Walt Taylor, Jack Ross and P. W. Harmon."

26. Test section can be considered the beginning of the present Quality Assurance Department in the Shipyard.

27. Linkletter biography of Wyler: "Wyler organized the Gyrocompass Shop and trained its personnel, devised test equipment provisions and quality control measures to produce complete overhauls of this instrument, so vital to ships."

28. The Metallurgical and Chemical Laboratories were combined under Mills in 1956. He retired following the consolidation of all the test laboratories in the Quality Assurance Division of the Production Department when the Shipyard began nuclear work.

29. Linkletter biography of Marriott: "Marriott led the development of radio work at the Navy Yard. He set up the Radio Laboratory and Shop, trained its personnel, supervised installations on shipboard and trained radio operators. He was one of the founders of the Institute of Radio Engineers."

30. SECRETARY OF THE NAVY REPORT, 1923, page 16 for details on aid to Japan.

31. Sand Point Archives, Entry 69-70691, report by Captain Gregory dated April 2, 1919.

32. The Marine Officers' Quarters are now Quarters S. The former Warrant Officers' Quarters are now H, I, J and K, west of Quarters R.

33. *SEATTLE TIMES*, June 29, 1920.

34. George R. Clark, A SHORT HISTORY OF THE UNITED STATES NAVY, pages 520-524. "The United States offered to scrap more of her ships than would be required of all the others together."

35. In April 1990, William P. Fisher of Ashland, Pennsylvania, wrote to the Bremerton Library requesting information about a 50-ton crane: *The crane was in operation alongside the pier where my ship, USS DETROIT, was tied up in the Bremerton Navy Yard for repairs during 1944. It was the largest steam crane I had ever seen, and I have never forgotten how it looked, newly repainted in black and red. Its well-oiled and greased condition bespoke the pride in its operation taken by the maintenance people and operators. I can still hear it chuffing quietly, doing its part for our war effort.* This crane, pictured on page 71, was dismantled in 1952.

36. Holden became a supervisor of new construction design. He retired in 1943 and died in 1981, aged 98. Page 58, second column, line 16 should read "1948."

37. It was not a happy time for Betty Fisher (Denschel) because she did not have playmates nearby.

38. Among these is Captain William T. Ross, who lived in the Yard when his father was Assistant Supply Officer during World War II. He was Executive Officer at the Naval Supply Center 1974-1975 and Comptroller at the Shipyard 1976-1979.

39. Noel Gayler graduated from the Naval Academy in 1935 and later became a four star admiral.

40. Ann Gayler's unpublished manuscript.

41. McKean was Senior Officer Present Afloat at that time.

42. Should read, "land was purchased from Helen Mc Chesney"

43. Line should read, ". . . (Bldg. 400) was erected in 1922. Arthur Thorson was swimming instructor at the pool for many years."

44. Some of the claimed bodies may have been reinterred at Bremerton's Ivy Green Cemetery on Naval Avenue. The southern part of this cemetery was the Charleston Cemetery, until the two towns consolidated.

45. Johnson was stationed at the hospital again in 1934, retired from the Navy in 1950 and died in Bremerton in 1988.

46. See page 214 for another photo of Dina Olsen.

47. "23nd" is how the word was written.

CHAPTER FOUR

1. USS LOUISVILLE — Commandant's memo of March 12, 1929 announced the naming of the 600-foot-long ship.

2. The Geneva Conference in July 1927 failed to reach any agreement on limitations on cruisers.

3. SALUTE, October 25, 1946. Figure is often rounded out to $8,600,000.

4. *SALUTE*, July 2, 1971.

5. Exact completion cost figures have not been found. SALUTE, October 25, 1946, says, "Two years and 27 days later LOUISVILLE was 85.1% completed, only $6,736,263 spent, $1,256,588 needed to finish the job."

6. Sand Point Archives Entry 74-70690. Religious services were held at Our Lady Star of the Sea Church with Father Joseph Camerman officiating. Military ceremonies were held at Bremerton Funeral Home.

7. In some copies of this photo the ship on the right is identified as USS PATOKA, an oil tanker equipped with a mooring mast and servicing facilities for dirigibles. However, all three aircraft carriers were at the Navy Yard in 1929, the date marked on this reproduction.

8. SECRETARY OF THE NAVY REPORT, 1927, page 624 said Seattle had one of the four Reserve Air Stations in the country.

9. Admiral James S. Russell, Float Planes and Flying Boats. Magazine and date unknown.

10. By 1936, a new ramp and hangar stood in the vicinity of the present battery shop, seaward of the Public Works Building.

11. According to Electrician Fred P. Johnson, who started work in the Electric Shop in 1923, electric-drive ships were very smooth going through the water but expensive to maintain.

12. *NAVAL HISTORY*, Fall 1990, pages 23-25.

13. Paquette (proper spelling) moved to Manette in 1925, was hired in 1947 by Olympic College to teach in the Apprentice program. He retired in 1980.

14. Nickname given during War of 1812 when enemy shot bounced off her sides.

15. Sand Point Archives. April 17, 1929, letter from the Commandant of the Thirteenth Naval District to the Secretary of the Navy and July 15, 1933, letter to the Chief of Naval Operations from USS CONSTITUTION's commanding officer Henry Harley about visit.

16. Bremerton and Charleston consolidated on January 1, 1928. Typewritten history of Charleston written by City of Bremerton Clerk Edward "Shine" McGowan.

17. July 15, 1933 H. Harley letter.

18. Captain Ernest Gayler had been recorder of the Board investigating the explosion. According to Mrs. Gayler's memoirs, "Ernest devised this method to separate storage far enough apart so that an explosion in one would not cause another to detonate. It was adapted by both services in the U. S. and in the French Army. Ernest was nicknamed Igloo Bill."

The Shipyard Chapel is decorated for Christmas, 1963. Except for the wall hangings, this beautiful chapel interior is today much the same as it was then. *Puget Sound Naval Shipyard*

19. Smith, homesick for the Northwest, played in the first game.

20. Moreell was appointed to USN Civil Engineer Corps in June 1917 and on December 1, 1938 became Chief of the Bureau of Yards and Docks. During World War II he organized the SeaBees, the Navy Construction Battalions. He was the only Civil Engineer in the U. S. Navy entitled to wear an Admiral's four stars.

21. PUGET SOUND NAVAL SHIPYARD, page 10 footnote.

22. H.G. Copley of the Forge Shop was one of the workers who came from Scotland.

23. Public Works records indicate Crane 28 is 125 feet high. A printed sheet labeled "General Information on Crane #28" says its purchase price was $498,496.55, and its per hour operating cost $110.

24. Published by the Master Mechanics and Foremen's Association.

25. One home and the guard station are now part of NAD Park off Kitsap Way. When land was being cleared in late 1970s to make room for more Jackson Park Housing, Dick Linkletter was able to have the other house and the stable considered for inclusion on the National Historic Sites Register. The Navy then gave the structures to Kitsap County and transported them to the present Waterfront Park in Old Town Silverdale, adjacent to the Kitsap County Historical Museum.

26. *SALUTE*, September 18, 1943, cites Central Tool Room Foreman W. E. Ainsworth's article in the October issue of the national trade publication, *THE TOOL ENGINEER*. Ainsworth was asked to write the article because "Centralized tool control and the wide extent in which it is practiced in the Yard has commanded considerable attention both on the part of other yards and private industry."

27. Building 443 is now used for offices. The Microcomputer Center and its teaching labs are located there as well as the Pass and Identification Office.

28. Handwritten order of May 12, 1937, signed by London, *SALUTE,* September 20, 1946, reported on Craven's visit to the Shipyard that month and said Craven was also responsible for remodeling of the old naval barracks at Manchester into a canteen and summer camping area for enlisted men. This was known as Camp Craven until the Navy took over the land shortly before World War II for use as a naval oil depot. The article also said Mrs. Craven was responsible for the Naval Hospital having a womens' ward.

29. *SALUTE*, November 7, 1969, says the Labor Board moved out of Building 56 into Building 157 prior to World War I, and later into Building 290.

30. Captain Gayler built the club in 1920 by selling $50 bonds to naval officers and Admiral Cottman's widow. He convinced a contractor to work for a lower price in order to keep his men busy; officers did much of the work themselves after hours.

31. A dogwood tree was planted in August 1988 in celebration of the Chapel's fiftieth anniversary.

32. INDUSTRIAL AND ADMINISTRATIVE HISTORY, page 89.

33. Spiller received an award in July 1959 for recommending design changes in turbo-generator system on USS IWO JIMA (LPH 2).

34. McIlraith designed a triple-walled shielded room for the Radio Shop.

35. Cowdrey was Design Superintendent 1935-1939. During World War II he served as assistant to Captain Homer Wallin at Pearl Harbor Navy Yard.

36. Johnson entered PSNY in 1917 as a copyist at $3.38 per day. He became Chief Engineer in the Design Division in May 1951, when Schairer retired.

37. Schairer entered PSNY in 1917 as a draftsman and became Design's Chief Engineer in 1943.

38. Most of these were recent University of Washington graduates including Richard Linkletter, William Spiller, and Stan Oliver.

39. This should have said "October 28, 1941".

40. Finch went to work at the Radio Material Office at PSNY in 1939. He left in 1944 to engage in Research and Development work in Los Angeles and for the Sandia Corporation.

41. See picture at bottom of page 97 for another radar system at PSNY.

42. Caldart, later Mayor of Poulsbo, also reported, "The ships company included a Royal Marine Band contingent. They were most impressive with their crisp precision, even when marching to waltz time up and down the dock alongside and in the general Yard area."

43. WARSPITE reunion in Bremerton is scheduled for May 1991.

44. *SALUTE* articles record other anniversary celebrations held after World War II.

CHAPTER FIVE

1. Alguard entered the Navy Yard in January 1935 as an apprentice machinist. He worked in various departments but considers his most important job was the ten years he spent as test engineer on steam catapults, before retiring in 1973.

2. They were working on the propellor of a tug needed to bring a ship into the Navy Yard on Monday morning.

3. Jim Downey started in the Design Division in 1941, but transferred to PERA(CV) in 1965 where he worked until retirement in 1974.

4. University of Washington graduate Linkletter was the first of the 84 junior engineers and naval architects to enter the Navy Yard's Design Division in 1938. He was selected as Tom Winson's Assistant Plant Electrical Engineer in 1941 and was a Weapons Systems Engineer in Ordnance Design from 1950 until his retirement in 1974.

5. Linkletter has written a seven-page paper on commuting from Seattle to Bremerton during the war years.

6. The tunnel was used for storage after World War II.

7. Puget Sound Navy Yard Demolition Plan was approved November 13, 1942, by Rear Admiral S. A. Taffinder.

8. Other junior officers stationed at the Navy Yard at that time, who later reached the rank of Rear Admiral, were Frank Jones, Jack Fee and Emery A. Grantham.

9. Grosso wrote a 15 page article on his experiences as a child in Puget Sound during World War II.

10. Schureman, president of the Kitsap County Historical Society 1979-1980, has written detailed accounts of life in Bremerton in the early 1940s.

11. "date that will live in infamy" is the wording President Roosevelt used, although it is usually quoted as "day that will live in infamy".

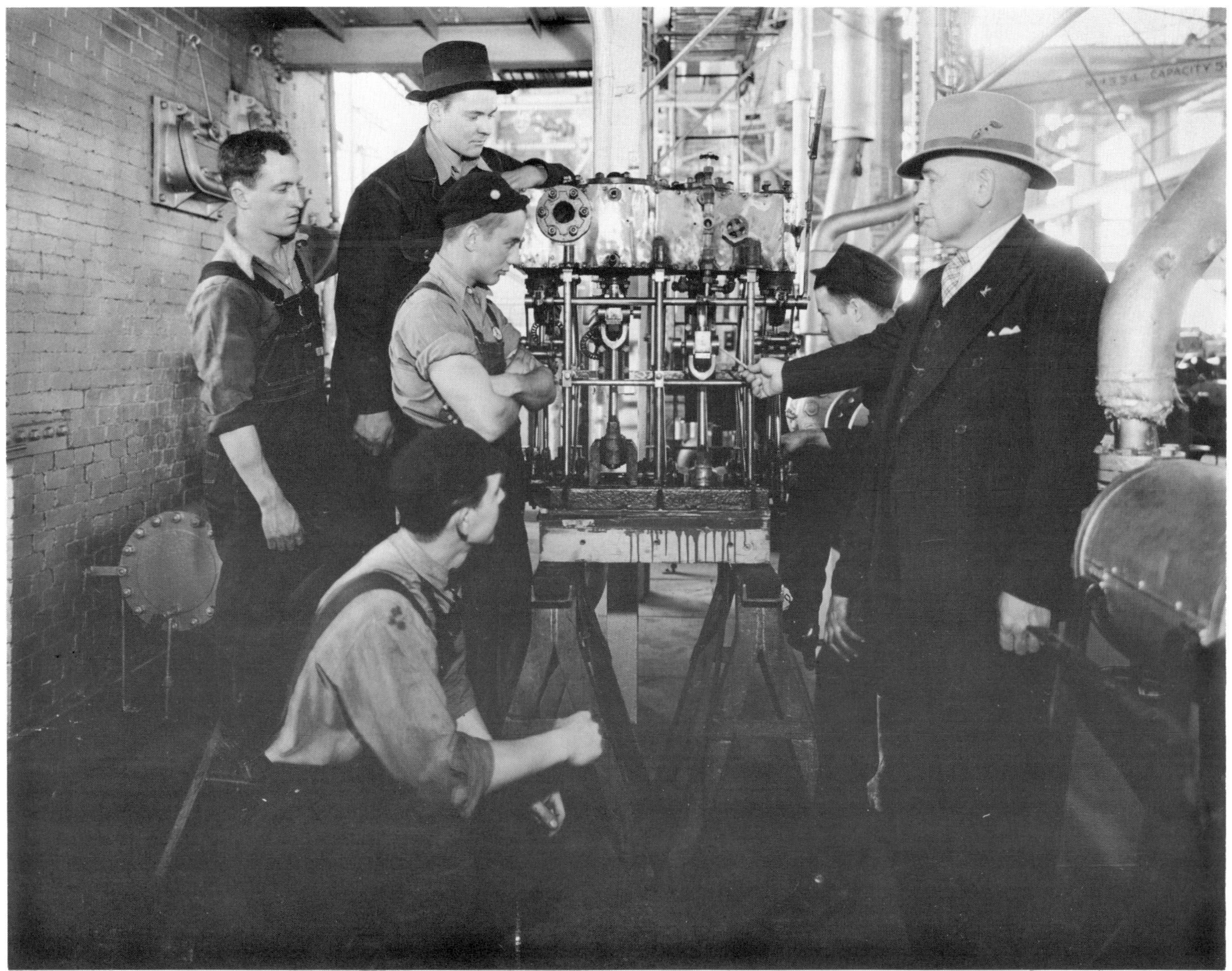

Because of the large increase in number of mechanics during World War II, training was of major importance. Here Theodore Peterson, Master of Shop 31, explains the workings of a reciprocating pump to his men. Peterson retired in August 1951.

Puget Sound Naval Shipyard

12. Fred Timmerman served in the Navy before entering PSNY in September 1939 as a helper rigger. He retired in February 1970 as Group Superintendent, Service Group.

13. The Japanese had reported that the battleships had been sunk.

14. The term "rifles" means guns mounted on ships.

15. On December 7, 1941, SARATOGA was enroute San Diego to pick up her planes.

16. Philip Sims, "Bulging Warships", *NAVAL ENGINEERS JOURNAL,* November 1989, pages 24-38. Blisters are external steel structures which provide void compartments on the side of a ship to provide torpedo protection. Blisters were proposed for SARATOGA and LEXINGTON as early as 1936; LEXINGTON was sunk before hers could be applied.

17. Johnson became a PSNY apprentice in 1923. In his memoirs he stated:" When we were told to do a job we took pride in doing the best we knew how and were glad we had a job."

18. Samuel Eliot Morison. HISTORY OF UNITED STATES NAVAL OPERATIONS IN WORLD WAR II, Volume Five, The Struggle for Guadalcanal, page 304.

19. The Japanese had reported her sunk.

20. PACIFIC NORTHWEST GOES TO WAR, pages 167-169.

21. Engineering officer Sprung had a distinguished career, retiring as a Captain with the Legion of Merit in 1953.

22. Philip Sims, "Bulging Warships", *NAVAL ENGINEERS JOURNAL,* November 1989, page 27. Sims says application of bulges (blisters) on battleships was used, among other things, to test the technique of welding.

23. Wynn entered the Navy Yard in 1940 as a helper in the paint shop. He became Quarterman Sandblaster in 1950 and Superintendent of the Paint Shop in 1977. He retired in 1980.

24. To ease crowded conditions on Farragut Avenue, Petersen moved the large assembled parts from the shop to the ships on barges.

25. Compiled from lists in INDUSTRIAL AND ADMINISTRATIVE HISTORY, BOOK II.

26. In the later days of World War II, Japanese pilots deliberately flew their planes into Allied ships to create as much damage as possible. Kamakaze is a Japanese word meaning "divine wind".

27. Letter in Kitsap County Historical Society files.

28. BDE stands for British Destroyer Escort.

29. Building ways replaced early seaplane landing area.

30. In this book a woman's married name is shown in parentheses, if she was married after the time mentioned.

31. Some examples of these models are at the Bremerton Naval Museum.

32. INDUSTRIAL AND ADMINISTRATIVE HISTORY, page 307.

33. West Park Housing was one of the first "garden apartment" complexes.

34. Lieutenant(jg) Frank J. Reh and Ensign Richard Scott were the officers on USS DEWEY.

35. The Josephs remained in Bremerton after the war.

36. Quincy Jr., attracted by music since early childhood, began playing with the encouragement of trumpet player, Eddie Lewis, proprietor of the Sinclair Heights barber shop. Jones' first performance as a professional trumpet player was at the Bremerton YMCA in 1947. He went on to star as composer, arranger, conductor and producer. The Jones family moved to Seattle in 1947. Lloyd Jones is active in the communications industry.

37. Lieutenant Commander Al Roos worked with Adams before Lieutenant(jg) Barnowe reported aboard.

38. The Print Shop.

39. Lawrence, who had been a carpenters mate in the Navy, spent over 25 years leading diving instruction and experimentation in the Yard before transferring to Washington D.C. as Ocean Engineering Operation Specialist in 1966.

40. Cooperative Asssociation was later known as Puget Sound Restaurant Association and the Employees' Services Restaurant Program.

41. Hope Beck, widow of State Representative C. W. "Red" Beck, worked at the Yard 1940-1973.

42. Doubleday entered the PSNY Shop Superintendents Office in 1940, was Administrative Assistant to the Planning Officer 1951-1963 and back in Production Department 1963-1966. He was in charge of recruiting in the Industrial Relations Office from 1966 until his retirement in 1970.

43. Burke, a 1937 University of Washington graduate in chemical engineering, worked for the Civil Service Commission as an examiner. He started at PSNY in 1940 where he became known as an expert on government labor procedures.

44. At that time most bench mark trade mechanics and supervisors were union members.

The replacement bow for USS TENNESSEE is moved west along Farragut Avenue toward Dry Dock 4 by 50-ton crane No. 34. The large building in the background is the Machine Shop. Puget Sound Naval Shipyard

45. Oliver entered PSNY in 1938 in the Electrical Group of the Design Section. He was elected secretary-treasurer of the Bremerton Metal Trades Council in January 1942. In 1948 he replaced Foster J. Pratt as president of the Technical Engineers and Draftsmen's Union in Washington, D.C. In 1957 he transferred to the Naval Ordnance Laboratory in White Oak, Maryland. During World War II, he was Bremerton Chairman of the Office of Price Control.

46. Renn entered the Yard in 1931 as a messenger, graduated from Apprentice School and became a tool maker. In 1942 he moved to the Shop Superintendent's Office as an engineer. His grandfather was a carpenter in the Yard 1898-1908, his father worked in the Navy Yard 1902-1943.

Kuhlman, who worked in Machinery Design, said the flotilla, "was started in 1942. The private boats were all painted Navy grey and the crews given naval rates. When the auxiliary's status was changed to temporary reserve, the owners of the boats were ranked as Lieutenant(jg). Most of the men in the flotilla were workers at PSNY but there were also others including Eddie Adams and Francis Shannon".

There are two misspellings in this paragraph. Correct spelling should be Ted Engebretson and Kermit Olson.

47. Unidentified newspaper clipping says, "PSNS employees invested $44,438,175 in bonds between November 16, 1941 and December 31, 1945.

48. Plating and Polishing, Pattern, Shipwright, Shipsmith, and Boiler, Building Trades and Sheetmetal Shops, the Captain of the Yard's Office, the Supply Department and the Disbursing Office.

49. Betty McGowan, who had worked in the Navy Yard since 1918, was chief clerk in the Shipwrights Shop. She was the daughter of early Bremerton contractor Michael F. McGowan, who helped construct many buildings in the Navy Yard and was active in community affairs.

51. According to William E. Irish, early in the war, ". . . if a day off fell on a weekday we were paid out of furlough pay. If it fell on Saturday or Sunday we received no pay for that day."

52. Alaska and Hawaii became states in 1959.

53. Masters was appointed Administrative Officer in the Industrial Relations Office in 1971.

54. Second line of this paragraph should read ". . . Blanchards and **Grandys**, to name . . ."

55. Flossie was laid off at the end of the war because her husband was working; her mother and brothers continued in government service.

56. Catherine Ward, the step-daughter of Dockmaster George Trahey, retired from the Design Division in 1955. Lulu Brown retired from the Industrial Relations Department in 1955.

57. Gladys Duley started work in the Navy Yard in 1913.

58. Fay Turpin was the wife of Chief Petty Officer "Dick" Turpin and Matron in Building 78.

59. Filion entered the Yard in 1927 and retired in 1962.

60. Restrictions against wearing wedding rings were relaxed for women who did not work around machinery.

61. Kitchen passed the bar in Oregon in 1937, worked 1942-1946 at PSNY in the Industrial Relations Office. After less than two years as a civilian she was commissioned as Lieutenant Commander. In 1962 she left the Navy and devoted herself to community affairs and oral histories.

62. The Association's Annual Shipyard Salmon Derbies began in 1947.

63. This was on Bob Hope's first USO tour to entertain workers and troops.

64. Nelle Klimas worked in Public Works before becoming Chief Clerk in the Captain of the Yard's office and then Administrative Assistant in the Naval Base Office.

65. In mid-1941, Secretary of the Navy Frank Knox recommended Navy installations start employee-oriented publications to provide information and boost morale. In January 1952 Reems introduced the first of six annual *SALUTE TO THE WORKER* issues with a larger size and pages in color. These featured the accomplishments of the Shipyard and its workers during the previous year.

66. Political columnist Ferguson's first byline was in *SALUTE.* In a column she wrote for the newspaper's fortieth anniversary, Ferguson described how Don Allyn helped her learn the ropes when he discovered her alleged previous newspaper experience was nonexistent.

67. *SALUTE,* August 28, 1943.

68. Letter from Secretary of the Navy to Commandant, Navy Yard Puget Sound, dated October 14, 1919.

69. Grenstad became a machinist apprentice in 1926 and entered the Director Shop in 1936. Retired 1963 as a Weapons System Engineer.

70. Building 455 is north of Dry Dock 6.

71. INDUSTRIAL AND ADMINISTRATIVE HISTORY, pages 438-439.

72. Every moving company in the Pacific Northwest was hired and the move was done over a weekend.

73. Sendner was only 29 when appointed master of the Foundry.

74. When Sendner retired in February 1944, his Chief Quarterman Molder, E. D. "Eddie" Boyle became Master of the Foundry. Boyle entered PSNY in 1916 as a "boy" in the Patternshop at $1.04 a day. Air conditioning, paved floor, and modern electric induction heating furnaces made his foundry one of the finest in the U.S. Boyle retired in 1955.

75. Wenker worked with the Civil Engineers while an enlisted mens' barracks was being built at the Ammunition Depot on Ostrich Bay. His name was given to the adjoining street which is now one of the major roads in Jackson Park Housing.

76. *SALUTE,* July 13, 1945.

77. Johnson retired from the Navy Yard in 1967.

78. Jones was a prolific writer on the staff of the Seattle Post-Intelligencer.

79. Glenn Whaley was the pilot who brought USS CUMMINGS to PSNS.

80. After being in the Navy four years, Smith entered the Navy as an electrician in December 1943. He retired as a Quarterman electrician. Smith is well known for his knowledge of ships.

81. According to Dorothy Germain Fiallo, her mother, Connie Osborne, who is pictured on page 142, was undoubtedly one who stayed.

CHAPTER SIX

1. SALUTE, November 9. 1945.

2. Later, retired Rear Admiral George A. Holderness Jr. was the representative in Washington, D. C.

3. Christie became Commandant of the Puget Sound Navy Yard and Commander of the Puget Sound Naval Base when Rear Admiral Robert Griffin became Commandant of the Thirteenth Naval District.

4. Indian Island is southeast of Port Townsend.

5. SALUTE, October 30, 1945.

6. Letter from Roland E. Perkins to Mayor Kean dated August 3, 1945.

7. SALUTE, January 31, 1947.

8. During World War II, Petersen's shop had 8,000 workers.

9. Peterson was one of the organizers of the Supervisors Club, now a chapter of the Federal Managers' Association.

10. The younger Hibbard started in the Public Works shops as laborer and surveyor and served in the Shop Superintendent's Office in World War I. He was active in the consolidation of Charleston and Bremerton. In 1940 Hibbard appointed a personnel supervisor in his shop, the first in any Navy Yard.

11. Penketh entered the Navy Yard in 1904 as a boilermaker first class, became Master of the Boiler shop in 1933 and retired in 1946. Herman Petersen told SALUTE in 1942 that Penketh was one of the best organizers in the business and the boiler shop one of the smoothest running operations in the Navy Yard.

12. McAfee entered the Yard in 1916 as a second class patternmaker and became master of the shop the day his predecessor, Hiram Richards, retired in 1938. McAfee retired in 1958 at which time he was the only man on the west coast to hold an honorary life membership in the American Foundrymen's Association.

Lloyd A. Prichard, the future Chief Planner and Estimator, presides at a Shipfitters and Welders Shops Chief Quarterman dinner in March 1946. Clockwise around the table from the left are: Ivan G. Smith, George Thomas, Richard Jacobs, Dean Calhoun, the future Structural Group Superintendent, Art Waaga, Bertie O. Gibbs, Prichard, Bill McDougall, Charles Lanning, James Sheriff, Carl Dick, Martin Sees and Wilfred Tucker. Puget Sound Naval Shipyard

13. Hicks entered PSNY in 1916 as a third class machinist. By the time he was 27 years old he was a leadingman and in 1951 was appointed Master of the Inside Machine Shop. E. S. Moskeland became shop master on Hicks' retirement in 1960.

14. Burzynski graduated fourth in his class at the U.S. Naval Academy. He retired in 1958.

15. Some of these models are on display at the Bremerton Naval Museum.

16. The Divers' Training School used equipment on a large barge. A platform secured ten feet off the bottom underneath Pier 4 provided a clear water working environment.

17. Other divers of this period were "Rosie" Evertson, T. R. Donahue, F. I. Tower, A. G. Hudson and J. H. Baines.

18. Diver's name is not known.

19. Salt water is a conductor of electrical current. Rubber coating on the metallic parts of diving equipment acted as an insulator.

20. On September 14, 1954, SALUTE quoted an Assistant Chief of the Bureau of Ships, "A total savings of $120, 911.60 has been realized from the first year's use of Formula 112."

21. The Director Shop was established to repair specialized equipment used in gun battery control systems. It was later expanded to include other optics and chronometers.

22. The Navy now uses quartz operated chronometers.

23. Hall entered the Shipyard in 1942 and was sent to the Aleutians to help the SeaBees install shore communication stations from Adak to Ketchikan.

24. Mason worked in RMO 1940-1950 at grades Radio Engineer P-1 to Electronics Engineer P-4.

25. Finch started work in RMO in the late 1930s. He left in 1944 to engage in Research and Development design "of ground control airplane approach and landing radar at Gilfillon in L. A."

Levin, who had been in the 1920 apprentice class, became head draftsman in RMO under C. E. Williams and then was Supervisor of Electronics Engineering in Design. Levin was well known for his sports activities and was elected into the Kitsap Sports Hall of Fame.

26. Campbell entered the Yard as an electrical/electronics apprentice in 1937. He worked in the Electric Shop, became the youngest leadingman in the yard and advanced to quarterman electrician and then quarterman radio-mechanic. After a tour in Planning and Estimating he became the Administrative Assistant in the Shop Superintendent's Office. He left the Shipyard in 1960 and retired in 1970.

27. The Bremerton Naval Museum has a photograph of President Truman entering the Shipyard where his attention was directed to a billboard showing PSNS' record on savings bond sales.

28. Buck Wynn, Personnel Safety Supervisor in the Paint Shop at this time, said sand blasters had been required to have annual x-rays at an earlier date and to wear masks and protective clothing designed by Katherine Kleist in the Sail Loft.

In September 1957 the Shipyard offered its workers the opportunity to receive Salk vaccine inoculation against polio.

29. Commercial artist Hegdahl joined the Training Section of the Industrial Relations Department in September 1947. He designed and did the technical work on billboards, SALUTE et cetera until his retirement in 1972.

30. Shipyard Riggers continued to assist with this project.

31. There is no sure way to keep fish from entering the dock. Noise is still used to chase fish from the dry dock before pumping. Large catches of fish no longer occur.

32. Sullivan, Commander of the Long Beach Naval Shipyard during its inactivation in 1949, instigated a vigorous job-hunting program for that Yard's displaced personnel.

33. Halligan had been Chief of the Bureau of Steam Engineering, Commanding Officer of one of the early airplane carriers, and Assistant to the Chief of Naval Operations before assuming command of PSNY on July 11, 1934. That winter he was operated on for a bleeding stomach ulcer and died December 11.

34. SALUTE, February 4, 1951.

35. The negotiating teams met at the 38th Parallel of latitude which became the de-facto boundary between North and South Korea.

36. In 1905 McLaughlin was Puget Sound Navy Yard's first apprentice joiner. He became Master of the Boat and Joiner Shop in June 1927, four years after the consolidation of the two shops.

37. Miles entered the section in 1941, worked with Fred Mills when the Metallurgy and Chemical Laboratories were combined and retired in 1969.

38. This was a team effort embracing many professions and trades in the development of a new process.

39. SALUTE, January 19, 1951.

40. Early in this period, Tool Shop Leadingman Gordon Munro received awards and a patent for the grinding wheel dresser improvements he had offered as a Beneficial Suggestion.

41. Beginning in 1963 the billet of Comptroller was held by Supply Corps Officers. In 1979 the billet reverted to line officers.

The caption under the picture on page 142 should list "Connie Osborne" not Osburn.

42. Max Josephson entered the Navy Yard in 1936 as a typist in the Inside Machine Shop, and started his career in the Accounting Department in 1938. He retired in 1970.

43. The only ceremony of its kind to be held on a U. S. warship.

44. Lindberg entered the Washington (D.C.) Navy Yard in 1900 as a machinist and transferred to PSNY in 1906. He retired because of illness in 1941 but returned to the Yard during the war to train supervisors.

45. Grenstad retired in 1963 as a Weapons System Engineer.

46. This was his pay as a journeyman.

47. SALUTE on September 14, 1954, listed the first thirteen apprentices: Art Holden, Stewart Chapman, Clifford Byrnes, Emil Olsen, Walter Lund, Walter G. Harris, Fritz Dahlquist, Oscar Martin, Herman Petersen, Ralph Siegner, W. J. Scott, L. W. Hill, and Louis Struck.

48. In 1991 the school is awaiting the appointment of a new director.

49. Olsen entered the Navy Yard in 1917 as a joiner. He transferred to Public Works in 1931 and was appointed master December 22, 1941.

50. Wallin came to PSNS from a tour as Chief of the Bureau of Ships. He had earlier been awarded the Distinguished Service Medal for his work in the salvage and repair of ships damaged in the Pearl Harbor attack.

CHAPTER SEVEN

1. Alaska-born Dolan was graduated from the Naval Academy in 1926. In 1954 he was advanced to the rank of Rear Admiral. Between his two tours as Commander of PSNS he was stationed in Washington, D.C. as Assistant Chief of the Bureau of Ships for Technical Logistics.

2. Frigate is an old Navy name for fast escort ships.

Looking east on Farragut Avenue in July 1955 is a Shipyard scene that was relatively unchanged for over three decades. The two officer's quarters at the top of "cardiac grade" look down over the familiar red brick shop buildings from the Machine Shop on the right, to the concrete structures of Building 290 and the Shipfitters Shop in the background.

Puget Sound Naval Shipyard

3. Mills started his sheetmetal shop apprenticeship in 1916 and became master of the shop in 1953. He retired in August 1958, at which time he was the Shipyard's senior employee.

4. The hull sections were butted together on the docking blocks and welders were placed at regular spaced intervals along the seam. They began welding in the same direction at the same time until a weld bead was completed all around the seam. This procedure was used for each of the multiple weld beads until the joint was complete. This welding technique minimized the built-in structural strains in a welded hull.

5. Fred Hicks, MAIN PROPULSION SHAFTING FOR THE GUIDED MISSILE FRIGATES USS COONTZ (DLG 9) AND USS KING (DLG 10). According to Captain D. R. Saveker, PSNS Assistant Repair Superintendent 1955-1958, credit is due Rear Admiral Charles Curtze, then Deputy Chief of the Bureau of Ships, who recognized the superiority of this method and authorized the funds necessary to do the tooling.

6. Alaska and Hawaii became the 49th and 50th states in 1959.

7. The USS HOEL (DDG 14) visited Seattle in 1988 with its GMLS Mark 11 system still in operation.

8. The Ordnance Design Subdivision was reorganized in 1988 and the Design Agency function work for the Mark 11 launcher was transferred to the Naval Ordnance Station, Louisville, Kentucky. The shaker table, which had vibration-tested the three-story high magazine and launcher system and other massive structures in Building 500 for almost thirty years, was scrapped.

9. LPH 1 was converted escort carrier BLOCK ISLAND (CVE-106).

10. Johnson presented a paper on AIRCRAFT CARRIER CONVERSIONS at the Military Engineers Conference in Seattle, Washington in 1955. He retired in April 1968.

11. Anti-submarine Rocket (missile).

12. Halvorsen entered the Navy Yard in 1934 as a helper pipefitter and began his apprenticeship the following year. In 1939 he started his career in the Design Division as an apprentice draftsman. He became Assistant Chief Design Engineer (Naval Architect) in 1956, a position he held until he retired in 1970.

13. A positive displacement gasoline storage and fueling system in which the volume of gasoline pumped from the inner tank was displaced by salt water in the outer connected tank. This was a safety measure which prevented gasoline vapors from forming in the tanks.

14. Finnegan entered the Navy Yard in 1917 as a draftsman and became a Marine Engineer in 1932. He and Carl Newstrom, head of Scientific Test branch, became GS-14s in 1958.

15. SALUTE on March 25, 1955, quoted Ship Superintendent Ray Burk (later a Rear Admiral) as calling the changed plans a "bombshell".

16. Controls aircraft on a carrier flight deck; comparable to the control tower at an airport.

17. Article not available.

18. Wortman started his machinist apprenticeship at the Navy Yard in 1940, became master of Shop 31 in 1968. Retired in May 1976, he is now a Commissioner on the Washington State Park Board.

19. These are aluminium alloy castings with special ductility characteristics.

20. Workman started as a messenger boy at the Navy Yard in 1927. He was involved in the introduction of computers into the Shipyard and appointed Supervisory Computer Specialist in 1968.

21. Distinguished Civil Service Award is the Navy's highest civilian award. Forsmark, Wesseler, Bruns and Clarence Peterson received the award in the mid 1950s.

22. Forsmark began his machinist apprenticeship in 1918 and became Shop 31 Foreman in 1952 when Fred Hicks became Master of the Machine Shop.

23. Wesseler entered the Navy Yard in 1923 as an Electric Shop apprentice, became Master of the shop in May 1944 and was later promoted to Electrical Group Superintendent. He retired in 1963.

24. Bruns started work in 1919 as a helper rigger.

25. Bench Mark trades are the basic trades that are used for comparing pay with trades in commercial enterprises.

26. Pins were also given for shorter periods of employment, starting at 5 years. The first 30-year pins at PSNS were presented to Helper Toolroom Mechanic Ray Kaness and Painter Charles J. Mapes in December 1949. Forty-year pins have a small ruby in the center.

27. After completing a business school course in Oregon, Genevieve Wolfe enlisted in the Navy Reserve as a Yeoman (F) and served for two years in the Supply Department at the Navy Yard, where she continued as a civilian after the war. Her first job title was "clerk" which was later changed to "typewriter."

28. Helen Miller returned to the Navy Yard in 1927, prior to which she had been secretary to the County Clerk of King County and a secretary in the office of the President of the University of Washington. She started in the Supply Department, soon transferred to Public Works and later became Secretary to the Public Works Officer.

29. Full name of the contractor was Manson, Jones, Perini and Osberg.

30. Internationally known plumbing and heating contractor Lents's was founded in Bremerton in 1907.

CHAPTER EIGHT

1. On July 7, 1986, SALUTE reported Admiral Rickover's death and said his first visit to Kitsap County had been in the 1920s as a junior officer aboard USS NEVADA.
2. Peterson remained head of the branch until his retirement in 1974, when Charles Henson took charge.
3. Charles Henson, named in the 1970 edition of OUTSTANDING YOUNG MEN IN AMERICA and given the Meritorious Civil Service Award in 1988 for his achievements in the field of ship silencing, wrote a six page article for NIPSIC TO NIMITZ regarding the Acoustic Range.
4. The system was operated and tested in underway transfer operations at sea for several years. PSNS made a number of changes to improve the system's operational reliability.
5. Boiler lighting-off ceremony indicates the ship is ready to generate steam to begin operating and testing the main machinery plant.
6. SIMON LAKE's speedy construction was necessary because Polaris-equipped submarines were nearing completion.
7. Executive Order 10988 established the government's Labor-Management Program in January 1962.
8. N. J. Repanich, later Mayor of Port Orchard, was the recorder for management for three labor agreements.
9. SALUTE June 26. 1964.
10. Roving is fiber glass matting.
11. Buher started as an apprentice electrician, transferred to Design and worked under Roy Levin, later transferred to Scientific and Test Section. On Marx Libby's retirement, Buher became the Radiation Control Coordinator.
12. As shown in movie made from Thomas Casey's book HUNT FOR RED OCTOBER.

This floating power plant is a steam generating plant that was used to supply steam to submarines to test their high pressure steam systems before lighting-off the ship's steam generators. The barge in the background is a living quarters barge used by submarine crews while their ship was not habitable during overhaul. Puget Sound Naval Shipyard

13. Olympic College and the University of Washington's College of Engineering offered background courses to round out the training program.

14. Kelly was married to Commandant Ziegemeier's daughter; his uncle, Captain Bruce Kelly, was Commanding Officer of the Bangor annex.

15. Lambert entered the Design Division in October 1940 and was head nuclear engineer in the Nuclear Power Divison before his retirement in 1973.

16. Campbell started his pipefitter apprenticeship in January 1942, served in the Navy from May 1944 until returning to the Pipe Shop in 1946. He transferred to Design in 1950 as engineering technician and became Chief Engineer in July 1972.

17. Emergency main ballast tank blow system (EMBT Blow) is a high volume system which by-passes the shipboard normal tank blow system in order to surface in an emergency.

18. Webber left PSNS in 1972 to become the Commanding Officer of the Mare Island Naval Shipyard. He was promoted to Vice Admiral in 1985, and served as Chief Engineer of the Navy until his retirement in 1987.

19. Humphrey wrote to the authors in March 1991, "In my opinion, facilities for nuclear work were better than at any other Shipyard. I was proud to be part of the nuclear power development at Puget Sound."

20. Two crews, complete with captain and other officers, are assigned to each SSBN. When the blue crew is at sea, the gold crew is in school or repairing and updating equipment and programs, and vice-versa.

21. 1967 Command History.

22. The Hydrofoil Unit was a satellite command at PSNS which received support services from the Yard.

23. Dry Dock 1 is the only dock at PSNS that is not deep enough to dock submarines without super flooding.

24. 1904 pictures show two trees in front of quarters C. The other tree, a Deodar Cedar, was removed during the 1930s.

25. A ceremony of historic significance to place a permanent monument at the site of the original flag raising.

26. Boring mill is still in place but plaque is covered.

27. Merrifield started his machinist apprenticeship in 1935 and became Foreman in the Machine Shop in 1958. He retired as Superintendent of Shop 31.

28. Cartwright entered the Shipyard in 1946 in Code 140. He transferred to the Production Department in 1960 as Production Engineering Superintendent, returning as Director of Code 140 in 1968. He retired in 1972.

29. 1967 Command History.

30. Information received from Captain Kevin McCook, Commanding Officer Naval Supply Center, Puget Sound.

31. 1965 Command History.

32. Known for many years as the Mariners' Club.

33. In 1986 the artifacts were moved to 130 Washington Avenue in Bremerton and set up as the Bremerton Naval Museum.

34. Formerly Bremerton Group, Pacific Reserve Fleet.

35. SALUTE, December 4,1970.

36. The Navy designated command of a shipyard as a captain's billet.

CHAPTER NINE

1. The move facilitated the construction of solid foundations for measurement equipment and helped ensure stable climate control. Production Department's Quality Assurance Division became a separate Department in 1966, with J. N. Wessel as Quality Assurance and Reliability Officer. Frank Reinhardt and Jean Livingston have written on the work of the Quality Assurance Laboratories.

2. Portal 3 is one of the four entrances from Farragut Avenue into the main tunnel, now part of Building 850.

3. Prior to August 1970 all calls into the Shipyard went through the switchboard.

4. One of these was George S. Callahan, the first president, who was City of Bremerton Housing Director during World War II. Callahan Drive on Bremerton's East Side was named for him. Others were Machinists P. P. Dolan and J. W. Fuller, Molder R. C. Faulkner and Electrician C. S. Oakley. Prichard later was Chief Planner and Estimator.

5. The hospital is two miles north of Bremerton on the east side of Highway 3, between the Austin Drive and Chico Way exits.

Summer Engineering Aides and Co-op Students in the Nuclear Engineering Department in 1981. Left to right: Brent Travis (ME-CalPoly), Tim Baltz (NE-UW), Laura King (IE-UC Berkeley), Jess Brown (IE-UC Berkeley), Julie Wurden (ChE-UW), Michael Griffith (CER-UW), Carol Ravano (CE-UC Berkeley), Kevin Hassett (ChE-WSU) and Ken Giffin (ME-UW).

Puget Sound Naval Shipyard

6. Part of the Marine command is Camp Wesley Harris, a marksmanship and tactical training site six miles south of the Bangor Submarine Base.

7. Bangor Naval Ammunition Depot was consolidated under Naval Torpedo Station as the Bangor Annex in 1970. Polaris Missile Facility, Pacific (POMFPAC), which had been established in 1964 as a tenant agency at Bangor, was recommissioned in 1974 as Strategic Weapons Facility, Pacific (SWFPAC). SUBASE BANGOR owns the railroad right of way between SUBASE and Gorst. PSNS owns the right of way from Gorst to Shelton.

8. The commanding officer's quarters were moved to Gig Harbor; other homes are at Lofall, Brownsville, Tracyton, Lone Rock and Port Orchard.

9. The Supervisors Association, now the Federal Managers' Association, was one of them.

10. When the Branch Clinic was built next to Building 850, the plants were transplanted to the hillside north of the Chief Petty Officers' Club.

11. The plaque was displayed in the lobby of Building 850.

12. Planners and Estimators are shop mechanics transferred to the Planning Department to estimate costs and write job orders.

The Combat Systems Office began with Thomas Tatham of the Director Shop in the late 1930s. As ordnance systems increased in quantity and complexity the cadre of highly skilled engineers increased. This included James Snell, Arie Vanderstaay, Denver Hackleman, Charles Titus, Elling Remmen, Wayne Stewart, Alvin Jensen, T. Gilbert, H. R. Brown, Frank Jackson, R. F. Anderson, D. C. Tufts, Ed Stamper, Ed Brady,

Red Beck and Mulvehill. Peter Schmickrath and John Volpone were Administrative Assistants to the Ordnance Officer. The Office is charged with weapon systems installation, check out, testing and storage handling capabilities. Its staff grew from 11 in 1957 to 82 in 1975.

13. Harriet McKiernan became Associate Editor in 1946. On transferring to the Keyport Torpedo Station in 1964 she started KEYNOTES, the station's newspaper. She returned to PSNS in 1966.

14. Herron retired in 1988 and Doreen Rekoski became editor. The editor in 1991 is Rita Kepner.

15. There had been other volunteer bands at the Yard. The Bremerton Symphony Orchestra was started during World War II as part of planned recreation for Navy Yard workers.

16. The Shipyard Employees Services Committee was established in 1965, taking over the activities of the Recreation Association which had earlier absorbed the Cooperative Association. The committee, of which Bradley Knight is the present chairman, is appointed by the Shipyard Commander and represents various civilian organization in the Shipyard. Operating with non-appropriated funds, Employees Services is charged with providing recreation, food services, shop and office welfare facilities and the negotiation of vending machine contracts.

17. These included Nort Atteridge, Gilbert Burt, Stanley Clarke, Dean Cook, C. B. "Tex" Dominy, Mike Gawenka, Herb Kimmel, Kaz Kimura, Joe Lambert, Jim Pappas, Claude Spigler, Roy Wright, Dick Albertson, Al Giannoti and others.

18. Pappas, a native Bremertonian, entered apprentice program in 1940. After serving a hitch in the Navy, he returned to the Shipyard. He became Superintendent of the Electric Shop in 1968 and retired in 1972.

19. Joe Valcauda, who was blind, successfully operated a newspaper and snack stand at the Main Gate for at least 20 years, starting in the late 1930s. He knew where everything was located in the stand his sister and niece stocked. He had no difficulty making change from coins; on the few occasions when he was given paper money, the Marines at the gate verified the amount.

20. Eagans entered the Yard in 1946 as a helper rigger. In June 1988 Kitsap County Commissioners officially changed the name of K Street to Loxie Eagans Way.

21. Engineering Librarian Sally Robinson in 1968, followed by Supervisory Personnel Assistant Vida Armstrong in 1970, held the job as collateral duty. On Pat Beatty's retirement in 1987, Ronnie Ocholik became Women's Program Manager.

22. Many Hispanics are employed in the Shipyard and are represented in the EEO Office.

23. Chief Petty Officer John Henry "Dick" Turpin, who worked as a rigger and diver in the Navy Yard following his retirement from the Navy, was a highly respected Black in the Bremerton area. Margaret Nieni described him as tall, erect and courtly. Hal Halvorsen remembered that even when in his 70s, Turpin could do handstands on a bollard. The Turpin Enlisted Men's Recreation Club at the Naval Air Station on Whidbey Island was named in his honor.

24. Barrios was instrumental in the establishment of Bataan Park at the intersection of Sylvan Way and Olympus Drive in East Bremerton.

25. A take-off on popular rental car advertisement.

26. The large frigates were all reclassified as cruisers.

27. See page 185. On turning command of the Shipyard over to the new Commander, Captain Jack W. Samford, Commodore Horne said: *The Shipyard would not be what it is today were it not for the people who have worked here through the years. It is their labor, vision and leadership that have brought us to this point . . . I have indeed been fortunate to work with such people.*

Horne's next duty station was at NAVSHIPS in Washington, D.C., where he was promoted to the two star rank of Rear Admiral.

CHAPTER TEN

1. USS BREMERTON (CA 130) was decommissioned in 1960 and stricken from the records in 1973.

2. As of April 30, 1991, USS NEW JERSEY is scheduled to return to PSNS for decommissioning.

3. Shores are timbers; strakes are longitudinal steel plates which make up the side of a ship's hull. The sheerstrake is the uppermost complete strake at the weather deck line of the hull.

4. Well remembered PSNS pilots include "Pappy" Beacham and Harry Owens.

5. ". . . takes the con" means takes responsibility for movement of the ship while underway.

6. SALUTE July 21,1978.

7. During docking and undocking operations, electric power and other utilities are disconnected from the ship, thereby halting work. For this reason docking operations are usually accomplished on the swing or graveyard shifts.

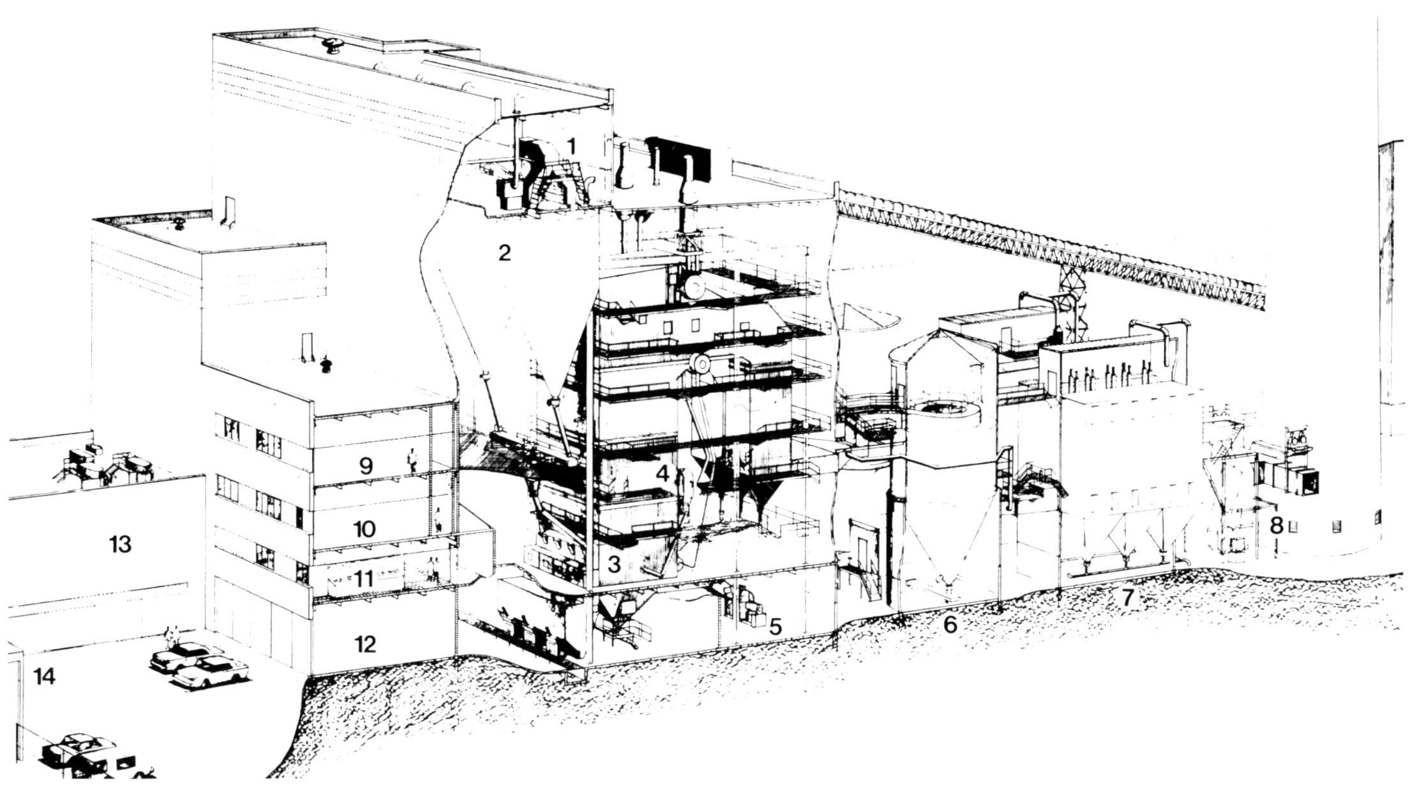

LOCATION OF MAJOR EQUIPMENT
(GENERAL FLOW PATH)

1 - Coal Handling System & Bunker Filling Traveling Tripper
2 - Twin Coal Bunkers for No. 2 Unit
3 - No. 2 Steam Generator Unit with Traveling Grate Stoker
4 - Grade No. 2 Fuel Oil Burners and Flame Safeguard
5 - Forced Draft Fan for Coal Burning Combustion Air
6 - Dry Scrubber Spray Absorber
7 - Fabric Filter Baghouse
8 - Induced Draft Fan and Stack

SHOPS AND OFFICES

9 - 4th FLOOR: Training, Engineering, Electrical Service Crew
10 - 3rd FLOOR: Shop 03 Superintendent, Supervision, Offices, Planning, Library, Electrical Maintenance Crew
11 - 2nd FLOOR: Dispatch Room, SCADA, Central Control Room, Shift Supervision, One-line Outages, Meter Room
12 - 1st FLOOR: Feedwater Room, Lab, Mechanical Distribution Crew, Wastewater Treatment Plant
13 - AIR COMPRESSORS, Shop Stores, Machine Shop, Pure Water Plant
14 - DIESEL HOUSE & SUBSTATION

Schematic drawing of the new Steam Plant. *Puget Sound Naval Shipyard*

8. SALUTE April 19, 1974.

9. A little known but vital operation of the Tool Shop is maintenance of sharp tools for ready distribution to waterfront mechanics.

10. The Planning Yard Division's function involves engineering and design work for ShipAlts (Ship Alterations) and OrdAlts (Ordnance Alterations) for the classes of Navy ships for which PSNS has planning yard responsibility. The Waterfront Division has engineering and design responsibility for ships at PSNS. The increased work load in both areas led to the split with a managing engineer for each division.

11. At the time this was considered to be sufficiently far outside the industrial area to be easy to keep clean.

13. Nelson was Commanding Officer of the hospital July 13, 1989 to March 29, 1991.

Page 227, The third line from the bottom in the right hand column should read "... are **pre**-World War II developments."

14. In 1940, 6,000 employees received instruction in one or more of 72 subjects in a training program set up by Shop Superintendent, Commander W. R. Dowd Sr.

15. Sam Josephson is the son of William "Bill" Josephson, who retired from PSNS in 1968.

16. From brochure entitled "The Power Plant".

17. Street was named for Marine Corps Lieutenant Eli K. Cole, who took charge of the Marine Barracks in 1899.

18. The second golf course, with its first and ninth holes near the officers' club,was closed when construction began on barracks for World War II.

19. Samuel Eliot Morison, THE STRUGGLE FOR GUADALCANAL, page 225.

20. VINSON's homeporting is only for the duration of her availability in the Shipyard.

21. Other shipyards have PERA responsibilities for other types of ships.

22. In May 1966 the name of the Bureau of Ships was changed to the Naval Ship Systems Command.

23. Move is expected to be made in August 1991.

24. After six years in the Navy, Pickard entered the Puget Sound Navy Yard in 1940 as a rigger working in the sail loft. He retired in 1973; the following year he returned to work as a "re-employed annuitant" in charge of planning in the Rigger Shop. Four months later, he was then selected as Production Department Quality Control Coordinator, working for Nuclear Production Manager Dale Estep.

25. Virginia Thorpe entered PSNY in April 1943, after working for the Department of Agriculture for five years. She had been secretary to seven Chief Design Engineers when she became receptionist at the Bremer/Wyckoff Building in 1988.

26. Pearline Giggans entered PSNY in September 1943 as an underclerk-typist, a CAF 1, earning $1,260 a year. In 1991 she is a Staff Assistant in IRO, a GS 7.

27. Wyckoff's name was submitted in 1983 by the Kitsap County Historical Society.

28. PSNS Planner and Estimator Rudy Campbell was Chairman of the Kitsap County Historical Society's Monument Project Committee when the granite marker was placed south of the sidewalk in 1951. Alice Wyckoff Swanson, Lieutenant Wyckoff's youngest daughter, unveiled the monument. In 1956 the Society sponsored the construction of a park in the area. The present park is on Shipyard land but is maintained by the Bremerton Park Department.

29. Keith also prepared the pattern for the memorial plaque in Bremer Park, east of Bremerton City Hall. Patternmakers are artists in wood carving who prepare wooden forms for patterns for castings.

30. Chapter 14 was organized in 1927 as the Supervisors' Club to promote cooperation between the various crafts in the Navy Yard. It became Local 14 of the National Association of Supervisors which had been organized in 1913 in Washington, D.C. The name was changed to the Federal Managers' Association in 1978, according to Aaron Tell, who represented the Association at its national conventions for 13 years.

31. The groups also are active in community affairs.

32. Major hostilities connected with Operation Desert Storm of the Gulf War had been completed.

33. THE SUN (formerly THE BREMERTON SUN) on April 28, 1991, announced the assessment the Navy had made of the value of its installations as a basis for recommending which bases should be closed to cut costs. Puget Sound Naval Shipyard was the only one of the eight naval shipyards to have the top rating in all categories: mission, land and facilities, contingency and mobilization potential and cost/manpower ratio.

ACKNOWLEDGEMENTS

Writing the definitive history of an organization as complex as the Puget Sound Naval Shipyard would be impossible without the help of many people. Our committee of five needed help gathering information, verifying facts, reviewing the manuscript and all of the other tasks associated with a project of this nature. Above all, we needed the collective memory of participants in the Shipyard's history to separate facts from myths.

Fortunately, many highly qualified men and women volunteered their services. Five of these people have worked with us since the beginning of the project. Dick Linkletter set the wheels in motion by soliciting memoirs from well over a hundred former Shipyard employees and provided information on a host of subjects. His wife, Doris, selected and copied innumerable articles from the Shipyard newspaper, SALUTE, organizing and entering the data on a computer.

Fred Timmerman, our source of information on Service Group Shops, sorted approximately 4,000 photos, comparing details and entering information on file cards. Fred, along with Ralph Smith, shared knowledge gained while serving in the Navy; Ralph's extensive library of ships' information and high-powered magnifying glass verified details in pictures to assure the accuracy of our ship information.

George Campbell gathered information from newspapers and people, and served as our "source person" on Planning Department matters. Others added their expertise and checked the manuscript for accuracy: Bob Doubleday, "Buck" Wynn, Harold Pickard and Bernie Schureman. Debbie Rollins and Mary Therese Bartlett proofread.

We wish we could include the names of everyone who helped. Suffice it to say, this book is the result of the work of many people who were anxious to tell the story of their shipyard.

While the book is primarily a birthday present to the Shipyard from the retired community, many others have played a part. We wish to thank the management of the Puget Sound Naval Shipyard for making records available for research, lending us photographs and providing a room for our material. Bob Essig and others in the Public Works Department's Planning Branch have researched dates and figures not available elsewhere.

Shipyard workers helped on their off-hours. Dave Bull transcribed taped interviews and copied old manuscripts. Linda Deline made posters for advertising this book as well as the book's art work, including the cover and title page. George Ganyon, Rhico Allen, Terry Nelson and Margo Gilbert helped with typing and errands.

We thank the Kitsap County Historical Society for agreeing to receive mail from book readers and Museum Curator Suzanne Anest for locating rare photographs.

Last, but not least, we wish to thank Chapter Fourteen of the Federal Managers' Association for financing the publication of this book, with special thanks to their book committee: Donnie Sprague, Rick Rollins, Nancy Shaffer, Steve Gunderson, Bill Fuhrmeister, Bob McClement, Sherie Lewis, Bill Billtoft, Donald Serry, Nancy Trask and Meredith Ames.

To those whom we have failed to acknowledge, here or on page 267, our sincere apologies and grateful thanks for your interest and help in producing this lasting legacy of the 100th anniversary of the Puget Sound Naval Shipyard.

Centennial Book Committee

Initial flooding of Dry Dock 6 before the dock's dedication on April 23, 1962, This picture, taken from the caisson, is a graphic display of turbulence resulting as water floods into the dock through the open valves in the caisson.

Grosso Collection, Kitsap County Historical Society

BIBLIOGRAHPY

BOOKS

Associated Editors, PACIFIC NORTHWEST GOES TO WAR, Art Ritchie and William J. Davis, Publishers, 1944

Bachiston, Marvin R., BATTLESHIPS, U.S. NAVY, Bend, Oregon, Maverick Publications, 1978

Clark, Rear Admiral George R., USN(Ret), A SHORT HISTORY OF THE U.S. NAVY, Philadelphia, J. B. Lippincott Co., 1939

Coontz, Admiral R. E., USN, FROM THE MISSISSIPPI TO THE SEA , Philadelphia, Dorrance & Co.Inc., 1930

Ewing, Steve, AMERICAN CRUISERS OF WORLD WAR II, Missoula, Pictorial Histories Publishing C., 1984

Fahey, James C., THE SHIPS AND AIRCRAFT OF THE U.S. FLEET , New York, GEMSCO Inc., 1944

Foss, William O., THE UNITED STATES NAVY IN HAMPTON ROADS , Norfolk, The Donning Co./Publishers, 1984

Friedman, Norman, U.S. CRUISERS, Annapolis, Naval Institute Press, 1987

Gibbons, Tony, THE COMPLETE ENCYCLOPEDIA OF BATTLESHIPS, New York, Crescent Books, 1983

Humble, Richard, BATTLESHIPS AND CRUISERS, Belgium, Basinghall Books Ltd., 1983

Humble, Richard, CARRIERS, Belgium, Bassinghall Book Ltd., 1982

Humble, Richard, SUBMARINES, Belgium, Basinghall Books Ltd., 1981

Janes, Francis P., R.N. Editor, JANES' FIGHTING SHIPS, New York, Franklin Watts, Inc., 1914-1919 (reprints)

Joint Task Force One Historian, OPERATION CROSSROADS, New York, Wm. H. Wise & Co., Inc., 1946

Kitsap County Historical Society Book Committee, KITSAP COUNTY HISTORY, BOOK III, Seattle, Dinner & Klein, 1977

Manette History Club, MANETTE PIONEERING, Portland, Taylor Publishing Co., 1988

Manning G. S. and Schumacher, T. L., PRINCIPLES OF WARSHIP CONSTRUCTION AND DAMAGE CONTROL, Annapolis, United States Naval Institute, 1935

McCurdy, H. W. MARINE HISTORY OF THE NORTHWEST, Seattle, Superior Publishing Co., 1966

McMurtrie, Francis E., Editor, JANES' FIGHTING SHIPS, New York, Arco Publishing Co.,Inc., 1944-45

McNeil, Jim, CHARLESTON'S NAVY YARD, Charleston, Naval Civilian Administrators Association, 1985

Miller, Nathan, THE U.S. NAVY, AN ILLUSTRATED HISTORY , Annapolis, United States Naval Institute, 1977

Moore, Captain John, R.N., Editor, JANES' FIGHTING SHIPS , New York, Franklin Watts, Inc.,1979-80

Morison and Rowe, SHIPS AND AIRCRAFT OF THE U.S. FLEET , Annapolis, Naval Institute Press, 1975

Morison, Samuel Eliot, HISTORY OF UNITED STATES NAVAL OPERATIONS IN WORLD WAR II, VOLUME FIVE, THE STRUGGLE FOR GUADALCANAL, AUGUST 1942-FEBRUARY 1943, Boston, Little, Brown and Company, 1949

Newell, Gordon and Smith, Vice Admiral Allen E., USN(Ret) , MIGHTY MO, New York, Bonanza Books, 1969

Polmar, Norman, SHIPS & AIRCRAFT OF THE U.S. FLEET, 12th edition Annapolis, Naval Institute Press, 1971

Potter, E. B., THE NAVAL ACADEMY ILLUSTRATED HISTORY OF THE UNITED STATES NAVY, New York, Thomas Y. Crowell Company, 1971

PUGET SOUND NAVAL SHIPYARD, Baton Rouge, Army & Navy Publishing Company, 1947

Reckner, James R., TEDDY ROOSEVELT'S GREAT WHITE FLEET , Annapolis, Naval Insitute Press, 1988

Rowe & Morison, SHIPS & AIRCRAFT OF THE U.S. FLEET, 9th edition, Annapolis, Naval Institute Press, 1971

Stillwell, Paul, Editor, AIR RAID: PEARL HARBOR!, Annapolis, Naval Institute Press, 1981

Swaney, Edwin S., OPERATION CROSSROADS, Montezuma, Iowa, Sutherland Publishing Co., Inc., 1986

Sweetman, Jack, AMERICAN NAVAL HISTORY, Annapolis, Naval Institute Press, 1984

United States Naval Academy Alumni Association, REGISTER OF ALUMNI, 1985 edition, Annapolis, Naval Institute Press, 1985

Wyckoff, Lieutenant A. B., USN, "Reminiscences of the Survey of Puget Sound and the Establishment of the Puget Sound Naval Station", THE WASHINGTON HISTORIAN, VOLUME 2, NO. 2. 1901

GOVERNMENT PUBLICATIONS

Hicks, Fred K., "Main Propulsion Shafting for the Guided Missile Frigates", Bremerton, Puget Sound Naval Shipyard, September, 1958

Mooney, James L. Editor, DICTIONARY OF AMERICAN FIGHTING SHIPS, 8 Volumes, Washington, D.C., Naval Historical Center, Department of the Navy, 1981

Report of Commission to select a site for a Navy Yard on the Pacific Coast north of the 42nd Parallel of North Latitude, 1889

Report of Commission to select a site for a Drydock on the Pacific Coast, 1890

Secretary of the Navy Annual Reports to the President

Wallin, Rear Admiral Homer, PEARL HARBOR: WHY, HOW, FLEET SALVAGE AND FINAL APPRAISAL, Washington, D.C. Naval History Division, Department of the Navy, 1968

Wortman, M.D. and Schroy, H., USS CONSTITUTION (CVA 66) REPAIRS, Puget Sound Naval Shipyard, 1971.

MAGAZINE ARTICLES

Fincher, Jack, "America's Deadly Rendezvous with the Spanish Lady" *SMITHSONIAN,* Washington, D.C., Thomas H. Black, January 1989

Payne, Steven A., "The Carrier That Lit Up Tacoma" *NAVAL HISTORY,* Annapolis, Naval Institute Press, Fall, 1990

Russell, Admiral James S.,USN (Ret), "Float Planes and Flying Boats", Magazine name, publisher and date not known

Sims, Phillip, "Bulging Warships" *NAVAL ENGINEERS JOURNAL,* Alexandria, American Society of Naval Engineers, November 1989

NEWSPAPERS

BREMERTON NEWS
BREMERTON SEARCHLIGHT
BREMERTON SEARCHLIGHT CONGRESSIONAL EDITION, December 1905
BREMERTON SUN
CHARLESTON RECORD
GLENWOOD HERALD
SALUTE
SEATTLE POST-INTELLIGENCER
SEATTLE TIMES
TACOMA NEWS TRIBUNE
THE SUN (Bremerton)

UNPUBLISHED MANUSCRIPTS, REPORTS, SCRAPBOOKS, JOURNALS AND MEMORANDA

Belote, William, HISTORY OF PUGET SOUND NAVAL SHIPYARD, 1982

Gavin, Lettie, AMERICAN WOMEN IN THE GREAT WAR

Grulich, HISTORICAL SURVEY OF PUGET SOUND NAVAL SHIPYARD (loose-leaf), Tacoma, Grulich Architecture & Planning Services, 1986

Holbrook, F.W.D. Scrapbook, 20 volumes of clippings from Puget Sound area newspapers.

INDUSTRIAL AND ADMINISTRATIVE HISTORY OF THE PUGET SOUND NAVY YARD 1891-1945.

Letters and Memorabilia read at Sand Point National Archives, Seattle, Washington, 1985-1986.

Schureman, Bernard, "Kitsap County, the War Years", 1980.

Talmadge, Charles E., "The Growth of the Puget Sound Naval Shipyard", Thesis for University of Washington, 1983.

University of Washington, Manuscript Collection.

University of Washington, Northwest Photograph Collection.

Wyckoff, Alice, Scrapbook of newspaper clippings.

Wyckoff, LT Ambrose B., JOURNAL.

MANUSCRIPTS AND ORAL HISTORIES MADE FOR NIPSIC TO NIMITZ

Contributors are:

Nort Atteridge, Jack Alguard, Maurice Allen, Vicente Barrios, Hope Beck, Roger Beckett, Walt Barry, Donald Buher, Orrin Burrows, Clyde Caldart, George Campbell, Theta Campbell, Gertrude Wall Carr, Steven Cartwright, H. G. Copley, Jim Downey, Ed Drouin, Phil Drouin, Bob Doubleday, Elbert Eliasen, William Filion, Douglas Finch, Richard Forde, Ann Gayler, Charles Gibson, John Gordon, Maurice Gowdey, Arvid Grenstad, Gerald Grosso, Harold Gunlach, Carlo Hallia, Halvor Halvorsen, Charles Henson, Janet Irvine, Gene Hoffman, Betty Henschel, Maryjayne Hladky, Gus Hudson, William Irish, Alonzo Johnson, Frances Johnson, Fred P. Johnson, George D. Johnson, Kenny Johnson, Max Josephson, Sam Josephson, Bertha Coontz Kokko, Hank Kuhlman, Earl Lagergren, Earl Lawrence, Dick Linkletter, Gertrude Madden, Betty McGowan, Frances McGowan, John Mason, Lynn Mickelson, Vic Mills, George Munro, Gordon Munro, Evelyn Holden Newkirk, Stan Oliver, Connie Osborne, Jim Pappas, Norma Pappas, Richard Parker, "Pete" Petrovic, Harold Pickard, Roger Paquette, Frank Reinhardt, Oliver Renn, Nick Repanich, Fred Rietson, Harold Richardson, Jack Ross, Dave Saveker, Mrs. C.E. Shepherd, Parker Snapp, Fred Timmerman, Adrian Vanderstaay, Oliver Vanderkleed, Arden Wilcox, Roy Workman, Mel Wortman, "Buck" Wynn, Chester Zukowski, Margo Zukowski.

APPENDICES

APPENDIX A

100 YEARS OF LEADERSHIP

In every organization there are a number of individuals who stand out from others because of their positions of responsibility. This section presents the Shipyard's lists of such persons.

Each of these persons has had a part to play in making the Yard's history, but there is not enough space in the book to cite even a single event for each. Because history takes place one day at a time, the names are presented in chronological sequence in each list.

An exception to this sequence is made in the lists of Shop and Group heads to account for an individual's promotion within those categories. The initial date shown is the date of that person's first appointment to a position which would qualify him to join the Master Mechanics and Foremen's Association, or its present day counter-part, the Superintendents' Association. The second date is the date of leaving the highest rate held by that person, either by transfer or by retirement.

The success of the Yard during its first century is due in large part to the work of these men. Their names stand as their memorial and as an inspiration to the men and women who will carry on the Shipyard's work in the second century.

COMMANDANTS AND SHIPYARD COMMANDERS

Lieutenant Ambrose B. Wyckoff	16 September 1891 - 1 February 1893
Commander J. C. Morong	6 February 1893 - 8 August 1896
Captain W. H. Whiting	8 August 1896 - 15 June 1897
Captain J. G. Green	19 June 1897 - 31 July 1899
Captain J. B. Coghlan	31 July 1899 - 2 July 1900
Captain W. T. Burwell	2 July 1900 - 29 August 1902
Rear Admiral Yates Stirling	29 August 1902 - 15 April 1903
Rear Admiral C. J. Barclay	1 July 1903 - 8 September 1905
Rear Admiral W. T. Burwell	11 September 1905 - 18 July 1908
Rear Admiral J. A. Rodgers	18 July 1908 - 26 July 1910
Rear Admiral V. L. Cottman	26 July 1910 - 14 February 1914
Commander D. W. Blamer (Acting)	15 February 1914 - 19 July 1915
Rear Admiral R. E. Coontz	20 July 1915 - 28 September 1918
Rear Admiral H. A. Field	28 September 1918 - 31 March 1921
Rear Admiral J. A. Hoogewerff	17 June 1921 - 1 October 1924
Rear Admiral J. V. Chase	1 October 1924 - 31 August 1926
Rear Admiral S. S. Robinson	1 October 1926 - 24 May 1928
Rear Admiral H. J. Ziegemeier	22 June 1928 - 15 October 1930
Rear Admiral E. H. Campbell	9 January 1931 - 12 May 1934
Rear Admiral John Halligan	11 July 1934 - 11 December 1934
Rear Admiral T. T. Craven	13 July 1935 - 8 May 1937
Rear Admiral E. B. Fenner	6 July 1937 - 30 August 1940
Rear Admiral C. S. Freeman	30 August 1940 - 31 March 1942
Rear Admiral S. A. Taffinder	31 March 1942 - 14 June 1944
Rear Admiral R. M. Griffin	15 June 1944 - 15 February 1945
Commodore W. M. Thompson	27 January 1945 - 19 December 1946
Rear Admiral E. W. Sylvester	30 December 1946 - 24 January 1950
Captain W. E. Sullivan	25 January 1950 - 4 May 1950
Rear Admiral H. E. Haven	17 May 1950 - 27 August 1953
Rear Admiral H. N. Wallin	27 August 1953 - 29 April 1955
Rear Admiral William A. Dolan	29 April 1955 - 5 June 1958
Rear Admiral Phillip W. Snyder	5 June 1958 - 27 May 1960
Rear Admiral William A. Dolan	27 May 1960 - 29 June 1962
Rear Admiral Floyd B. Schultz	29 June 1962 - 15 March 1967
Rear Admiral William F. Petrovic	15 March 1967 - 30 June 1972
Captain F. F. Manganaro	30 June 1972 - 5 June 1976
Captain J. K. Nunneley	5 June 1976 - 16 June 1979

Captain J. H. Boyd Jr.	16 June 1979 - 20 June 1981
Captain R. B. Horne	20 June 1981 - 25 August 1984
Captain J. W. Samford	25 August 1984 - 24 June 1988
Captain Arthur Clark	24 June 1988 - Present

PLANNING OFFICERS

Captain R. B. Hilliard	19 September 1927 - 15 October 1928
Captain L. S. Border	22 January 1929 - 1 June 1931
Captain C. L. Brand	20 May 1931 - 16 June 1932
Commander R. T. Hanson	6 August 1932 - 26 August 1936
Commander B. S. Bullard	6 September 1936 - 28 February 1939
Commander R. S. Hitchcock	31 March 1939 - 12 January 1943
Captain W. A. Brooks	6 January 1943 - 4 December 1944
Captain H. E. Haven	1 February 1945 - 22 August 1947
Captain Logan McKee	10 July 1947 - 14 March 1949
Captain W. P. Webster	14 March 1949 - 28 February 1951
Captain H. J. Pfingstag	28 February 1951 - 31 May 1952
Captain J. B. Duval Jr.	1 June 1952 - 11 July 1954
Captain C. E. Trescott	12 July 1954 - 12 May 1955
Captain P. W. Pfingstag	13 May 1955 - 17 January 1956
Captain R. L. Mohan	18 January 1956 - 2 January 1958
Captain W. F. Cassidy	20 February 1958 - 3 December 1958
Captain P. F. Shetenhelm	4 December 1958 - 19 January 1960
Captain J. W. Dolan Jr.	20 January 1960 - 31 July 1960
Captain D. K. Ela	30 August 1960 - 7 August 1962
Captain F. A. Spencer	7 August 1962 - 12 June 1963
Captain H. A. Jackson	13 June 1963 - 10 July 1964
Captain J. R. Newland	10 July 1964 - 7 June 1966
Captain W. A. Yatch	8 June 1966 - 19 August 1968
Captain E. R. Meyer	19 August 1968 - 30 September 1970
Captain J. H. Webber	1 October 1970 - 14 July 1972
Captain R. K. Reed	9 August 1972 - 30 June 1974
Captain W. A. Lent	1 July 1974 - 12 July 1977
Captain A. E. Keegan	13 July 1977 - 30 June 1980
Captain H. C. Hunter	1 July 1980 - 24 June 1983
Captain R. W. Hoag II	24 June 1983 - 19 June 1985
Captain M. Nickelsburg	19 June 1985 - 6 June 1988
Captain A. W. Hunt	7 June 1988 - 22 September 1989
Captain B. A. J. Baumann	23 September 1989 - Present

CHIEF DESIGN ENGINEERS

Schairer, Henry T.	1943 - 1951
Johnson, Ruben E.	1951 - 1954
Johnson, Kenneth G.	1954 - April 1968
Newstrom, Carl A.	April 1968 - July 1971
Chamberlin, Carl H.	July 1971 - July 1972
Campbell, George G.	July 1972 - April 1981
LaFountaine, Robert L. (Planning Yard)	April 1981 - Present
Anderson, L. M. (Waterfront)	1987 - Present

PRODUCTION SUPERINTENDENTS

Captain A. T. Church	16 August 1921 - 31 January 1922
Commander R. L. Irvine	12 February 1922 - 14 April 1926

CHIEF PRODUCTION OFFICERS

Captain H. G. Bowen	8 October 1926 - 1 July 1931
Captain C. A. Dunn	1 July 1931 - 1 February 1934

PRODUCTION OFFICERS

Captain H. M. Cooley	2 April 1934 - 25 May 1935
Captain C. A. Bonvillian	17 June 1935 - 5 June 1936
Captain F. J. Wille	2 July 1936 - 21 February 1940
Captain P. M. Dunbar	4 March 1940 - 18 February 1942
Captain P. W. Hains	23 February 1942 - 2 February 1943
Captain F. M. Earle	2 February 1943 - 15 February 1943
Captain P. B. Nibecker	7 March 1943 - 21 November 1946
Captain L. A. Kniskern	21 November 1946 - 14 April 1949
Captain R. C. Bell	15 April 1949 - 28 February 1951
Captain H. P. Webster	1 March 1951 - 31 May 1952
Captain H. J. Pfingstag	1 June 1952 - 13 July 1954
Captain J. V. Duval Jr.	13 July 1954 - 18 May 1955
Captain C. E. Trescott	18 May 1955 - 31 August 1955
Captain P. W. Pfingstag	27 January 1956 - 7 June 1957
Captain J. J. Fee	17 June 1957 - 29 October 1958
Captain W. F. Cassidy	29 October 1958 - 22 January 1960
Captain P. E. Shetenhelm	23 January 1960 - 15 November 1960
Captain R. J. Nesbitt	16 November 1960 - 13 June 1962
Captain J. F. Ellis Jr.	10 September 1962 - 16 June 1966
Captain J. R. Newland	16 June 1966 - 29 June 1967
Captain E. R. Meyer	30 June 1967 - 18 August 1968
Captain F. J. Reh	19 August 1968 - 31 December 1972
Captain W. L. Martin	31 December 1972 - 6 August 1976
Captain J. I. Webb	6 August 1976 - 31 July 1978
Captain T. A. Marnane	1 August 1978 - 9 July 1980
Captain L. C. Gies	9 July 1980 - 4 June 1982
Captain J. W. Samford	4 June 1982 - 25 August 1984
Captain R. E. Traister	25 August 1984 - 31 December 1985
Captain J. L. Taylor	31 December 1985 - 5 September 1989
Captain P. F. Scardigno	5 September 1989 - Present

PRODUCTION DEPARTMENT SHOP HEADS

Note: An "X" preceding the shop number indicates a direct expense shop; "O" indicates an overhead expense shop.

MASTER MECHANICS

Trahey, George W.	Shipwright X64	1904 - 6 June 1936
Stewart, Robert	Joiner C & R	1904 - 1927
Wolfkill, S. G.	Machine C & R	1904 - 1926
Hibbard, R. E.	Sheetmetal X17	12 December 1911 - 28 February 1934

Shop heads (and others) taken in August 1904. Left to right, upper row — George W. Trahey, Master Shipwright; Harry Bailey, Chief Draughtsman; Thomas Raines, Quartermaster Painter in Charge; Timothy Twiggs, Warrant Carpenter; C. A. Douglas, Foreman of Laborers. Lower row — A. L. Croxton, Master Electrician; David Rogers, Master Shipfitter; J. H. Warren, Quartermaster Blacksmith in Charge; Robert Stewart, Master Joiner; A. J. Stanton, Leading Boatbuilder; S. G. Wolfkill, Master Machinist; E. W. Nelson, Quartermaster Molder in Charge. Puget Sound Naval Shipyard

Richards, H.	Pattern X94	1912 - 30 May 1938
Rogers, David	Shipfitter X11	1912 - 1919
Stephenson, Bart	Boiler X41	1912 - 1932
Gruwell, F. M.	Shipfitter X11	1913 - *
Gregoire, George	Shipfitter X11	1919 - August 1923
Smith, F. A.	Machine X31	1919 - 1936
Petersen, Herman	Shipfitter X11	16 December 1919 - 31 August 1948
Gruwell, F. A.	Pipe X56	1920 - 1934
Fryette, R. B.	Electric X51	May 1920 - June 1943
Jaixen, H. C.	Paint X71	24 January 1921 - 8 August 1946
Sender, J. L.	Foundry X81/94	7 January 1922 - 19 February 1943
Abbott, W. E.	Marine Machine X38	1 January 1926 - 5 August 1941
McLaughlin, C. V.	Boat X64	June 1927 - 2 February 1951
Hauschel, Julius	Pipe X56	1934 - 1940
Penketh, George W.	Boiler X41	1 January 1935 - 31 January 1946
Farrell, King	Forge X23	16 May 1935 - 20 September 1942
Wagner, Ernest H.	Electric X51	29 July 1935 - May 1944
Hibbard Russell J.	Sheetmetal X17	23 March 1936 - 30 April 1953
Peterson, Theodore	Machine X31	11 May 1936 - 31 August 1951
McArthur, A.	Shipwright X61	15 June 1936 - 31 January 1945
McAfee, E. J.	Pattern X94	1 June 1938 - 31 January 1958
Boone, Earl	Shipfitter X11	1940 - 31 July 1954
Sheldon, Gail	Pipe X56	16 September 1940 - 31 August 1950
Ainsworth, W. G.	Tool O6	1941 - 28 February 1955

Hicks, Fred K.	Machine X31	1941 - 29 January 1960
Jolly, C.	Marine Machine X38	3 March 1941 - 30 June 1947
Chandler, J. B.	Pipe X56	1942 - 30 April 1957
Kettel, J. M.	Forge X23	7 December 1942 - 30 June 1951
Mills, Victor L.	Sheetmetal X17	August 1943 - 1 August 1958
Boyle, Ed	Foundry X81/94	17 January 1944 - 31 March 1955
Bruns, Walter H.	Rigger X72	29 May 1944 - 31 July 1960
Hilberg Victor V.	Woodworker X64	20 December 1944 - 24 April 1962
Nelson, A. W.	Marine Machine X38	23 July 1945 - 31 July 1960
Personett, A.	Paint X71	26 August 1946 - 31 August 1950
Scmickrath, Anthony	Boiler X41	26 August 1946 - 31 December 1951
Cooley, James S.	Paint X71	1 January 1951 - 20 September 1962
Pendras, R. P.	Boiler X41	3 March 1952 - 30 December 1965

Master Mechanics and Foremen in a group picture taken in May 1943. Left to right. Front row — C. V. McLaughlin, Master Boatbuilder; Theo Peterson, Master Machinist; Harry Stoner, Foreman Shipfitter, loft; Herman Petersen Master Shipfitter; J. L. Sendner, Master Molder; George W. Penketh, Master Boilermaker; Henry C. Jaixen, Master Painter; William Hensel, Foreman Shipfitter, welder; Herman Boldt, Foreman Mechanic, Naval Torpedo Station, Keyport. Middle row — Lee P. Allison, Foreman Machinist, gyro and instrument; Virgil W. Driver, Master Mechanic, N. A. D.; Arthur J. Lent, Foreman Pipefitter; Earl C. Jolly, Master Machinist, outside; Russell J. Hibbard, Master Sheetmetal Worker; Edward J. McAfee, Master Patternmaker; Ernest H. Wagner, Master Electrician; Christ F. Rasmussen, Foreman Sailmaker; Walter H. Bruns, Foreman Rigger; Fred Hicks, Foreman Machinist. Back row — Oliver I. Olsen, Master Mechanic, Public Works; Albert E. Taylor, Foreman Sheetmetal Worker; Earle H. Boone, Foreman Shipfitter; Gale E. Sheldon, Master Pipefitter; Clarence H. Macomber, Master Mechanic, transportation; Andrew McArthur, Master Shipwright; Jack Kettel, Master Forge Shop; Walter E. Ainsworth, Foreman Toolmaker; Riley Masterson, Master Mechanic, power plant; and Lynn Mickelson, Foreman Public Works. Puget Sound Naval Shipyard

SHOP FOREMEN

Douglas, C. A.	Laborer C & R	1904 - *
Bankhead, William	Power Plant	1911 - *
Chrey, Ted	Shipfitter X11	1940 - 24 March 1950
Rasmussen, Chris	Sailmaker X74	9 October 1940 - 1943

Allison, L. P.	Machine X31	14 October 1940 - 30 April 1952
Stoner, Harry	Shipfitter X11	1941 - *
Taylor, A. E.	Sheetmetal X17	1941 - 30 September 1955
Campbell, James	Laborer X72	1942 - 15 September 1961
Lent, A. J.	Pipe X56	1942 - *
McKinney, Kent	Shipfitter X11	1942 - *
Willow, Elmer	Shipfitter X11	1942 - *
Hensel, W. J.	Welder X26	7 September 1942 - 25 October 1961
Higgins, F. P.	Electric X51	5 July 1944 - 3 September 1959
King, Earl R.	Rigger X72	11 June 1948 - 21 July 1950
Hall, Thomas C.	Electronics X67	23 August 1948 - 31 March 1967
Arthun, C. T.	Marine Machine X38	17 February 1950 - 31 January 1956
Cameron, George C.	Forge X23	2 July 1951 - 21 April 1964
Hill, George M.	Rigger X72	7 April 1952 - 27 October 1959
Richards, Charles	Pipe X56	7 April 1952 - 31 December 1959
Forsmark, Carl A.	Machine X31	15 August 1952 - 30 September 1957
Waaga, Art	Shipfitter X11	27 September 1954 - 15 March 1964
Bay, Curtis	Pipe X56	6 August 1959 - 8 July 1966
Bert, John	Welder X26	7 March 1961 - 7 May 1962
Woodings, H. K.	Woodworker X64	29 April 1962 - 15 February 1967

GROUP SUPERINTENDENTS

Wesseler, William O.	Electrical 950	7 June 1943 - 31 May 1963
Calhoun, Dean C.	Structural 920	2 August 1954 - 31 May 1963
Moskeland, E. S.	Machinery 930	28 March 1955 - 30 August 1963
Tennyson, E. H.	Outfitting 940	19 December 1955 - 31 July 1964
Cummins, James L.	Machinery 930	24 October 1958 - 31 December 1974
Timmerman, Fred L.	Service 970	8 January 1960 - 24 February 1970
Berg, Berge A.	Electrical 950	10 June 1962 - 30 April 1968
Longmate, J. R.	Structural 920	23 June 1963 - 7 April 1978
Miles, Ray M.	Electrical 950	18 December 1966 - 29 June 1973
Ensign, Jack B.	Service 970	31 December 1967 - 27 December 1974
Tower, Frank	Service 970	8 May 1970 - 8 October 1976
Wixson, M. S.	Structural 920	7 March 1971 - 3 July 1987
Morgan, H.	Structural 920	2 May 1971 - 29 December 1981
Holt, George H.	Electrical 950	25 June 1972 - 12 January 1979
Edwards, E. R.	Electrical 950	23 December 1973 - 21 December 1981
Westermann, Leo H.	Machinery 930	24 June 1974 - 29 June 1984
Dotson, Robert L.	Service 970	4 July 1976 - Present
Woodward, Gary W.	Electrical 950	9 May 1979 - Present
Kelly, Owen A.	Machinery 930	13 July 1980 - July 1989
Main, Jerry	Pipe/Boiler 960	July 1985 - Present
Burton, A. L.	Structural 920	25 May 1986 - Present
Frey, John	Structural 920	5 July 1987 - 24 June 1988
Barrington, Ron	Mechanical 930	June 1989 - Present

SHOP SUPERINTENDENTS

Langhans, John	Foundry/Pattern X81/94	10 September 1955 - 8 October 1976
Edwards, Colin	Pipe X56	25 November 1957 - 31 March 1972
Merrifield, A. J.	Machine X31	28 April 1958 - 31 July 1970
Emery, C. A.	Sheetmetal X17	9 June 1959 - 29 June 1973
Trull, James C.	Paint X71	29 April 1962 - 10 September 1971
Wakefield, P. W.	Rigger X72	29 April 1962 - 11 September 1968
Thomas, S. G.	Welder X26	7 May 1962 - 25 October 1966
Bjorling, C. A.	Electric X51	5 July 1963 - 25 July 1966
Eslick, A. W.	Marine Machine X38	12 December 1963 - 13 July 1970

O'Neill, E. J.	Shipfitter X11	27 September 1964 - 13 December 1970
May, Merle M.	Tool O6	20 December 1964 - 12 July 1972
Nelson, Kenneth B.	Boiler X41	24 October 1965 - 16 May 1971
Shroy, Harold M.	Welder X26	19 June 1966 - 22 October 1972
Turner, J. L.	Woodworker X64	24 August 1966 - 10 March 1971
Wortman, Mel D.	Machine X31	4 January 1968 - 14 May 1976
Pappas, James T.	Electric X51	25 August 1968 - 23 June 1972
Kimmel, J. T.	Electronics X67	3 October 1968 - 21 December 1973
Trower, Herbert O.	Woodworker X64	2 March 1970 - 26 June 1972
Smith, Thomas L.	Marine Machine X38	7 June 1970 - 29 June 1973
Potter, O. J.	Paint X71	9 August 1970 - 25 February 1977
McCaughan, B.	Shipfitter X11	24 January 1971 - 31 July 1975
Laurie, Bruce	Pipe X56	2 April 1972 - 28 March 1975
Kidrick, E. F.	Woodworker X64	25 June 1972 - 22 December 1977
Oass, J. P.	Sheetmetal X17	25 June 1972 - 30 December 1976
Melz, R. L.	Tool O6	9 July 1972 - 29 February 1980

This picture of the Centennial Superintendents was taken in March 1991; all Superintendents were present. Left to right. Back row — Utilities Superintendent Gary Champine, Electronics Superintendent Gerald Phipps, Shipfitter Superintendent Deryl George, Maintenance Superintendent Robert Fojtik, Boiler Shop Superintendent Jerry Locke, Tool Shop Superintendent Roy Reece, Outside Machine Shop Superintendent Arthur Fenton, Electrical Superintendent Floyd Sawyer and Insulator Superintendent James Davidson. Middle row — Welder Superintendent Mark Banks, Machinist Superintendent Tom Hamilton, Public Works Group Superintendent Mickey Hall, Sheetmetal Superintendent Ronald Rust, Painter Superintendent Mark Matthews, Transportation Superintendent Norbert Hermanson, Forge/-Foundry/ Pattern Superintendent Glenn Anderson and Rigger Superintendent Robert Henderson. Front — Machinery Group Superintendent Ronald Barrington, Structural Group Superintendent Albert Burton, Woodcrafter Superintendent Lewis Commons, Pipefitter Superintendent James Backs, Service Group Superintendent Robert Dotson, Boiler/Pipe Group Superintendent Jerry Main and Electrical Group Superintendent Gary Woodward.

Puget Sound Naval Shipyard

Owen, D. L.	Marine Machine X38	8 July 1973 - 18 July 1980
Gurley, C. W.	Temporary Service X99	30 September 1973 - Present
Whiton, J. E.	Electronics X67	2 January 1974 - 19 December 1981
Chandler, M. T.	Pipe X56	15 February 1976 - 9 July 1982
Thompson, W. L.	Shipfitter X11	15 February 1976 - 25 May 1977
Drouin, E. A.	Machine X31	16 September 1976 - 11 January 1980
Wynn, Everett R.	Paint X71	8 May 1977 - 11 January 1980
Carlson, W. B.	Shipfitter X11	22 May 1977 - 1 September 1978
Heyer, Willard L,	Rigger X72	6 September 1977 - 28 October 1983
Gibson, C. H.	Pattern X94	11 September 1977 - 30 December 1982
Shelgren, A. E.	Boiler X41	29 September 1978 - 2 January 1987
Yoder, D.	Shipfitter X11	29 September 1978 - 16 August 1982

Christenbury, W.W.	Woodworker X64	11 November 1978 - 28 December 1981
Worms, T. A.	Machine X31	10 February 1980 - October 1989
Nyswonger, D. R.	Paint X71	9 March 1980 - 15 August 1987
Armstrong, L. F.	Sheetmetal X17	22 April 1980 - 3 June 1986
McNellis, T. L.	Tool O6	4 May 1980 - May 1989
Phipps, G. H.	Electronics X67	21 February 1982 - Present
David, G. A.	Electric X51	10 March 1982 - 30 October 1987
Chamberlin, E.	Pipe X56	9 July 1982 - 12 April 1986
Ross, Lanny	Welder X26	1 September 1982 - 2 December 1989
Commons, L. E.	Woodworker X64	5 September 1982 - Present
Forbes, J.	Shipfitter X11	5 September 1982 - 31 October 1986
Sherman, D. M.	Rigger X72	30 October 1983 - 1 December 1988
Bradley, R.	Forge/Foundry X23/81	December 1983 - May 1987
Williams, C. J.	Marine Machine X38	23 September 1984 - 13 May 1990
Locke, J. E.	Boiler X41	February 1986 - Present
Brockner, P.	Shipfitter X11	20 July 1986 - 30 December 1989
George, Deryl S.	Shipfitter X11	20 July 1987 - Present
Mannen, H. Keith	Paint X71	12 August 1987 - 1 March 1991
Backs, J. A.	Pipe X56	October 1987 - Present
Rogers, F. S.	Forge/Foundry X23/81	10 September 1988 - July 1989
Isham, L.	Rigger X72	12 December 1988 - 31 August 1990
Rust, Ron	Sheetmetal X17	June 1989 - Present
Anderson, Glenn	Forge/Foundry X23/81	August 1989 - Present
Mathews, Mark	Paint X71	27 February 1991 - Present
Hamilton, Tom	Machine X31	September 1989 - Present
Reece, Roy	Tool O6	April 1990 - Present
Sawyer, F. R.	Electric X51	30 April 1990 - Present
Davidson, James	Pipe X57	May 1990 - Present
Fenton, Arthur R.	Marine Machine X38	14 May 1990 - Present
Banks, Mark	Welder X26	June 1990 - Present
Henderson, Robert	Rigger X72	2 September 1990 - Present

PUBLIC WORKS OFFICERS

Civil Engineer T. C. McCollom	September 1891 - November 1892
Civil Engineer U. S. G. White	November 1892 - May 1894
Civil Engineer R. C. Hollyday	May 1894 - April 1897
Civil Engineer F. O. Maxson	May 1897 - March 1899
F. W. D. Holbrook	March 1899 - February 1903
Civil Engineer A. C. Lewerenz	February 1903 - November 1907
Civil Engineer P. L. Reed	December 1907 - September 1910
Civil Engineer E. H. Brownell	September 1910 - July 1913
Captain(CEC) L. E. Gregory	July 1913 - May 1920
Captain(CEC) E. R. Gayler	October 1920 - May 1924
Captain(CEC) W. H. Allen	November 1924 - September 1926
Captain(CEC) E. R. Gayler	October 1926 - May 1930
Lieutenant Commander(CEC) B. Moreell	June 1930 - May 1932
Captain(CEC) R. M. Warfield	July 1932 - October 1936
Captain(CEC) E. R. Gayler	October 1936 - December 1938
Captain(CEC) R. E. Thomas	December 1938 - December 1941
Captain(CEC) E. Phillips	December 1941 - November 1942
Captain(CEC) R. C. Harding	November 1942 - October 1943
Captain(CEC) E. B. Keating	October 1943 - December 1947
Captain(CEC) H. W. Johnson	December 1947 - October 1949
Captain(CEC) R. P. Carlson	October 1949 - March 1952
Commander(CEC) L. W. Reeder	March 1952 - March 1954
Captain(CEC) F. L. Endebrock	March 1954 - September 1957

Captain(CEC) P. W. Boothe	September 1957 - July 1960
Captain(CEC) J. J. Albers	July 1960 - May 1962
Captain(CEC) E. I. Mosher	May 1962 - January 1964
Captain(CEC) D. K. Culp	February 1964 - April 1966
Captain(CEC) R. C. Jensen	April 1966 - January 1970
Captain(CEC) A. W. Lalande	January 1970 - October 1971
Captain(CEC) G. W. Schley	October 1971 - August 1974
Commander(CEC) C. R. Williams Jr.	August 1974 - September 1974
Captain(CEC) R. A. Bafus	September 1974 - May 1976
Captain(CEC) C. P. Anderson	May 1976 - July 1979
Captain(CEC) J. A. Westcott	July 1979 - June 1983
Captain(CEC) Q. E. D. Lewis	June 1983 - September 1986
Captain(CEC) R. F. Heine Jr.	September 1986 - September 1989
Captain(CEC) T. J. Tanner	October 1989 - Present

PUBLIC WORKS DEPARTMENT SHOP HEADS

MASTER MECHANICS

Croxton, A. L.	Power Plant	1904 - 1934
Macomber, C. H.	Transportation O2	1913 - 31 May 1943
Warren, J. H.	Utilities O3	1913 - 1933
Parker, M.	Utilities O3	October 1932 - May 1934

Master Mechanics and Foremen in February 1966. Left to right. Top row — Public Works Group Master L. A. Welch, Foreman Pipefitter C. B. Bay, Foreman Inside Machinist A. J. Merrifield, Foreman Sheetmetal Worker C. A. Emery, Electrical Group Master B. A. Berg, and Foreman R. Hentschel, NAD, Bangor. Middle row — Foreman Shipfitter E. J. O'Neill, Service Group Master F. L. Timmerman, Foreman Electronics Mechanic T. C. Hall, Foreman Painter J. C. Trull, Master Maintenance Mechanic C. A. Addy and Foreman Woodworker H. K. Woodings. Front row — Machinery Group Master J. L. Cummins, Master Pipefitter C. W. Edwards, Foreman Rigger P. W. Wakefield, Foreman Electrician C. A. Bjorling, Master Molder J. E. Langhans, Foreman Boilermaker K. B. Nelson, Foreman Machinist F. E. Davies, NTS, Keyport, and Foreman Toolmaker M. M. May. Absent when the photo was taken were: Structural Group Master J. F. Longmate, Foreman Welder S. G. Thomas, Foreman Machinist A. W. Eslick, Utilities Foreman T. W. Maughan and Transportation Foreman B. D. Hunter. Puget Sound Naval Shipyard

Olson, O. I.	Maintenance O7	9 August 1941 - 30 September 1957
Masterson, R. C.	Utilities O2	April 1942 - August 1945
Montain, G. R.	Transportation O2	7 June 1943 - 27 January 1946
Thompson, H. D.	Utilities O3	6 August 1945 - 27 January 1961

SHOP FOREMEN

Jones, E. T.	Maintenance O7	1929 - 1941
Mickelson, Lynn	Maintenance O7	April 1942 - April 1954
Wooldridge, C. E.	Maintenance O7	14 June 1954 - 28 February 1965

GROUP SUPERINTENDENTS

Welch, L. A.	Public Works 450	28 October 1946 - 30 June 1967
Mauchan, T. W.	Public Works 450	29 January 1961 - 13 April 1971
Laurie, Cyril C.	Public Works 450	31 December 1967 - 29 June 1979
Prentice, Deane	Public Works 450	16 March 1973 - 15 October 1982
Hall, Mickey	Public Works 450	June 1980 - Present

SHOP SUPERINTENDENTS

Addy, C. A.	Maintenance 07	17 March 1958 - 2 March 1966
Lund, T. L.	Utilities O3	12 May 1963 - 30 July 1965
Hunter, B. D.	Transportation O2	30 January 1966 - 28 May 1971
Rasmussen, G. S.	Utilities O3	2 December 1967 - 23 March 1973
Paschal, L. C.	Maintenance O7	17 May 1970 - 10 September 1976
Stansbury, W.	Transportation O2	1 November 1970 - 10 January 1976
Sandin, E.	Transportation O2	7 December 1975 - 29 May 1981
Werner, H. L.	Maintenance O7	1 May 1977 - 31 May 1985
Hermanson, N.	Transportation O2	29 May 1981 - Present
Lyons, E. J.	Utilities O3	20 March 1983 - 1 August 1988
Champine, G. S.	Utilities O3	July 1988 - Present
Fojtik, Bob	Maintenance O7	October 1989 - Present

*An * indicates that the dates are not known.*

SUPPLY OFFICERS

P.A.Paymaster E. B. Webster	15 June 1892 - 16 April 1894
Paymaster J. R. Martin	17 April 1894 - 5 January 1895
P.A.Paymaster H. R. Sullivan	1 January 1895 - 10 May 1897
Pay Director W. W. Williams	20 April 1897 - 1 June 1898
Asst. Paymaster G. Brown Jr.	1 June 1898 - 24 May 1899
Asst. Paymaster E. C. Tobey	24 May 1899 - 1 October 1900
Pay Inspector W. J. Thompson	1 October 1900 - 10 January 1901
Asst. Paymaster C. J. Peoples	11 December 1900 - 15 March 1901
Paymaster H. A. Dent	15 March 1901 - 17 December 1901
P.A.Paymaster C. Morris Jr.	1 November 1901 - 6 February 1903
P.A.Paymaster D. M. Addison	31 January 1903 - 28 February 1906
Paymaster H. R. Insley	28 February 1906 - 30 April 1906
Paymaster H. H. Balthis	30 April 1906 - 30 June 1906
Paymaster J. Brooks	30 June 1906 - 2 January 1908
Paymaster R. Spear	2 January 1908 - 22 August 1910
Paymaster G. Brown Jr.	22 August 1910 - 27 April 1914
Pay Inspector T. S. O'Leary	1 May 1914 - 28 March 1917
Paymaster E. T. Hoopes	20 March 1917 - 21 July 1919
Commander(SC) S. E. Barber	15 July 1919 - 6 February 1922
Commander(SC) J. F. Hatch	12 January 1922 - 23 February 1925
Lieutenant Commander(SC) R. S. Robertson	23 February 1925 - 30 March 1925

Captain(SC) A. F. Huntington	30 March 1925 - 1 June 1929
Captain(SC) H. de F. Mel	1 June 1929 - 29 May 1933
Commander(SC) O. D. Conger	30 May 1933 - 31 August 1933
Captain(SC) J. F. Hatch	1 September 1933 - 21 March 1935
Commander(SC) H. C. Gwynne	22 March 1935 - 30 April 1935
Captain(SC) C. Conard	1 May 1935 - 15 August 1935
Captain(SC) B. M. Dobson	17 September 1935 - 30 June 1930
Captain(SC) W. S. Zane	1 July 1939 - 20 September 1942
Captain(SC) S. J. Brune	21 September 1942 - 17 August 1944
Captain(SC) H. M. Shaffer	18 August 1944 - 22 May 1947
Captain(SC) J. J. Jecklin	23 May 1947 - 3 August 1949
Captain(SC) B. D. Smith	4 August 1949 - 18 July 1952
Commander(SC) W. H. Schleef	19 July 1952 - 27 August 1952
Captain(SC) W. R. Wright	28 August 1952 - 13 September 1954
Captain(SC) J. C. DeWitt	14 September 1954 - 28 January 1957
Captain(SC) M. G. Stokes	29 January 1957 - 30 July 1958
Captain(SC) E. W. Sutherling	31 July 1958 - 5 August 1960
Captain(SC) G. C. Heffner	6 August 1960 - 23 July 1963
Captain(SC) H. H. Hunt	23 July 1963 - 30 November 1965
Captain(SC) D. Keers	1 December 1965 - 5 July 1968
Captain(SC) C. E. Whiteside	6 July 1968 - 31 July 1972
Captain(SC) E. M. Wieseke	1 August 1972 - 28 June 1974
Captain(SC) D. M. Carpenter	29 June 1974 - 23 September 1977
Captain(SC) J. C. Ardizzone	23 September 1977 - 31 August 1981
Captain(SC) R. A. Lingenbrink	1 September 1981 - 14 September 1984
Captain(SC) D. K. Anderson	15 September 1984 - 25 July 1990
Captain(SC) R. W. Rogers	25 July 1990 - Present

ACCOUNTING OFFICERS

Lieutenant Commander(SC) M. C. Shirley	1920 - 1922
Commander(SC) C. V. McCarty	1922 - 1926
Commander(SC) J. T. Hatcher	1926 - 1928
Captain(SC) R. W. Johnston	1928 - 1929
Captain(SC) T. P. Ballenger	1929 - 1930
Captain(SC) W. H. Wilterdink	1930 - 1933
Captain(SC) J. R. Hornberger	1933 - 1934
Captain(SC) J. E. McDonald	1934 - 1936
Captain(SC) E. C. Morsell	1936 - 1937
Commander(SC) C. V. McCarty	1937 - 1939
Commander(SC) A. G. King	1939 - 1942

FISCAL OFFICERS

Captain(SC) S. E. Smith	September 1942 - November 1945
Captain(SC) C. Schraaf	November 1945 - June 1946
Captain(SC) H. E. McMahan	June 1946 - January 1947
Commander(SC) C. J. Lee	January 1947 - July 1947
Captain(SC) M. W. Willard	July 1947 - August 1950
Commander(SC) T. W. Ragland	August 1950 - December 1952
Commander(SC) R. W. Oliver	December 1952 - July 1953

COMPTROLLERS

Captain M. H. Gluntz	July 1953 - May 1954
Captain J. B. Rawlings	May 1954 - January 1957
Captain(SC) J. J. Payton Jr.	January 1957 - July 1957
Captain J. W. Dolan Jr.	July 1957 - January 1960

Captain W. M. Nicholson	January 1960 - July 1962
R. A. Finke (Acting)	July 1962 - July 1963
Captain(SC) R. B. Edwards	July 1963 - July 1965
Captain(SC) J. L. Graham	July 1965 - January 1968
Captain(SC) G. W. Gallagher Jr.	January 1968 - July 1972
Captain(SC) R. E. Graham	August 1972 - June 1974
Captain(SC) P. B. Crouch	June 1974 - April 1975
Captain(SC) W. T. Ross Jr.	April 1975 - July 1979
Captain J. E. Smith	August 1979 - July 1985
Captain W. J. Metzger Jr.	August 1985 - August 1990
Captain T. E. Bugarin	August 1990 - Present

ORDNANCE OFFICERS

Captain F. C. Stelter Jr.	1 April 1946 - 3 February 1949
Captain C. F. Piper	29 January 1949 - 30 January 1950
Commander E. V. Bruchez	31 January 1950 - 12 May 1952
Commander A. R. McWilliams Jr.	13 May 1952 - 14 August 1952
Commander O. D. Compton	15 August 1952 - 27 March 1955

ORDNANCE SUPERINTENDENTS

Commander O. D. Compton	28 March 1955 - 5 August 1955
Commander D. V. Hickey	5 August 1955 - 15 August 1958
Captain M. M. Gantar	15 August 1958 - 29 April 1960
Commander R. W. Scott	12 August 1960 - 1 July 1962

COMBAT SYSTEMS SUPERINTENDENTS

Commander R. W. Scott	1 July 1962 - 22 July 1963
Commander J. L. Chelgren	22 July 1963 - 23 June 1967
Lieutenant Commander H. G. Morris	23 June 1967 - 19 May 1969
Commander W. B. Abbott	28 July 1969 - 4 October 1971

COMBAT SYSTEMS OFFICERS

Commander W. B. Abbott	4 October 1971 - 30 June 1972
Commander C. M. McCarty	5 September 1972 - 2 February 1975
Lieutenant Commander H. J. Holliday	3 February 1975 - 9 June 1977
Lieutenant Commander H. L. Thompson	10 June 1977 - 9 August 1977
Commander H. M. Effron	10 August 1977 - 8 April 1980
Commander W. G. Paulsen	15 July 1980 - 22 July 1981
Lieutenant Commander W. H. Moore	23 July 1981 - 14 September 1984
Commander T. N. Chance	15 September 1984 - 30 June 1987
Commander D. J. Grandia	16 November 1987 - 15 August 1988
Commander D. J. Armstrong	24 October 1988 - Present

COMMANDING OFFICERS, NAVAL SUPPLY CENTER

Captain(SC) S. M. Ball	2 October 1967 - 12 September 1969
Captain(SC) S. R. Simpson	13 September 1969 - 30 August 1972
Captain(SC) F. J. Woodworth	31 August 1972 - 30 April 1974
Captain(SC) G. J. Thompson	1 May 1974 - 15 April 1976
Captain(SC) F. H. Kela	16 April 1976 - 25 August 1978
Captain(SC) W. E. Lindsay	26 August 1978 - 20 June 1980
Captain(SC) E. K. Walker	21 June 1980 - 31 July 1981
Captain(SC) A. H. Allnutt	1 August 1981 - 29 July 1983
Captain(SC) C. C. Klein	30 July 1983 - 31 July 1985

Captain(SC) W. S. Draper, IV	1 August 1985 - 24 July 1987
Captain(SC) R. W. Fenick	25 July 1987 - 13 June 1989
Captain(SC) K. W. McCook	14 June 1989 - Present

COMMANDING OFFICERS, NAVAL HOSPITAL

Passed Asst. Surgeon J. Stroughton	7 April 1895 - 23 November 1895
Assistant Surgeon C. D. Brownell	23 November 1895 - 10 December 1896
Passed Asst. Surgeon F. W. Olcott	18 February 1897 - 11 January 1898
Passed Asst. Surgeon A. Farenholt	26 January 1898 - 31 March 1898
Surgeon A. G. Cabell	7 April 1898 - 19 September 1898
Assistant Surgeon H. C. Curl	25 September 1898 - 23 January 1899
Assistant Surgeon C. R. Burr	31 January 1899 - 16 October 1900
Passed Asst. Surgeon L. L. VonWedekind	16 October 1900 - 4 November 1901
Passed Asst. Surgeon A. W. Dunbar	4 November 1901 - 23 November 1902
Passed Asst. Surgeon A. R. Alfred	23 November 1902 - 9 December 1904
Surgeon D. N. Carpenter	9 December 1904 - 5 July 1906
Surgeon J. B. Dennis	5 July 1906 - 2 February 1908
Surgeon C. B. Bagg	2 February 1908 - 21 March 1910
Medical Director H. E. Ames	28 March 1910 - 24 December 1910
Surgeon A. Farenholt	21 November 1910 - 8 April 1911
Surgeon F. C. Cook	10 May 1911 - 8 April 1914
Surgeon E. M. Shipp	8 April 1914 - 17 June 1914
Medical Director A. R. Wentworth	1 July 1914 - 20 November 1918
Captain(MC) L. L. Von Wedekind	20 November 1918 - 8 October 1919
Captain(MC) T. H. Berryhill	30 October 1919 - 1 June 1920
Commander(MC) C. C. Grieve	1 June 1920 - 7 July 1921
Captain(MC) C. D. Kindleberger	20 July 1921 - 7 July 1924
Captain(MC) T. W. Richards	7 July 1924 - 13 July 1927
Rear Admiral(MC) N. J. Blackwood	13 July 1924 - 5 December 1929
Captain(MC) U. R. Webb	5 December 1929 - 31 October 1931
Captain(MC) F. E. Porter	31 October 1931 - 15 February 1935
Captain(MC) D. C. Cather	15 February 1935 - 30 September 1939
Captain(MC) A. L. Clifton	30 September 1939 - 4 December 1941
Captain(MC) D. Hunt	4 December 1941 - 4 January 1943
Captain(MC) E. D. McMorries	4 January 1943 - 11 July 1945
Captain(MC) J. R. Fulton	11 July 1945 - 12 August 1946
Captain(MC) J. H. Robbins	12 August 1946 - 29 September 1947
Captain(MC) S. S. Cook	4 November 1947 - 28 March 1949
Captain(MC) M. C. Mathis	28 March 1949 - 30 September 1951
Captain(MC) C. G. Clegg	30 September 1951 - 1 October 1953
Captain(MC) T. D. Boaz	11 October 1953 - 20 February 1956
Captain(MC) E. T. Knowles	20 February 1956 - 30 March 1957
Captain(MC) C. R. Moon	30 March 1957 - 16 July 1959
Captain(MC) R. L. Ware	16 July 1959 - 31 January 1961
Captain(MC) E. P. McLarney	31 January 1961 - 30 September 1964
Captain(MC) J. E. Gorman	30 September 1964 - 30 June 1967
Captain(MC) C. P. Root	30 June 1967 - 30 June 1969
Captain(MC) R. G. Brown	30 June 1969 - 19 July 1971
Captain(MC) W. T. Lineberry Jr.	19 July 1971 - 3 June 1974
Captain(MC) H. E.Glick	3 June 1974 - 1 July 1974
Captain(MC) H. P. Pariser	1 July 1974 - 19 July 1977
Captain(MC) R. C. Elliott	19 July 1977 - 23 June 1980
Captain(MC) R. A. Proulx	11 July 1980 - 2 September 1983
Captain(MC) L. E. Nelson	2 September 1983 - 6 June 1985
Captain(MC) J. W. Winebright	6 June 1985 - 28 July 1987
Captain(MSC) J. W. Bartlett	28 July 1987 - 13 July 1989
Captain(MC) R. A. Nelson	13 July 1989 - 29 March 1991
Captain(MSC) R. E. McKee	29 March 1991 - Present

APPENDIX B

SHIPYARD FAMILIES

This information was submitted by the person identified with *. If the information came from another source, that source is identified. The date shown is the approximate start of work date.
** Information not available.

FIVE GENERATION FAMILIES

Henry Karst Father	1903 Laborer, Paint Shop
Holt Karst Son	1917 Laborer
C. J. (Jim) Karst* Grandson	1940 Electronics Mechanic
Susan Karst Pace Great-granddaughter	1973 Engineering Technician
Darrell Krouse Great-great-grandson	1989 Apprentice Electronics Mechanic
Lissa Krouse Great-great-granddaughter	1989 Apprentice Painter

---

Benjamin Franklin Tabbutt Father	circa 1893 Worked on Dry Dock 1
Amos Tabbutt Son	** Shipwright
Julian Beck Grandson	1923 Pattern Shop
Doris Beck Garringer Watson* Great-granddaughter	1951 Management Analyst
David E. Garringer Great-great-grandson	Electrician

FOUR GENERATION FAMILIES

Milton Albert Campbell Father	1911 Foreman Shipfitter
Rutherford Campbell Son	1914 Estimator
Milton P. Campbell Grandson	1947 Design
Jack M. Campbell Great-grandson	1976 Rigger Shop

Submitted by Loretta Campbell, widow of Milton P.

Jay Cheatham Father	1941 Maintenance Shop
Betty Cheatham Spigler* Daughter	1942 Paint Shop
Edward Jay Spigler Grandson	1966 Welder Shop
Sheila M. Spigler Great-granddaughter	1991 Rigger Shop

John Dick Father	1903 Boiler Shop
Carl Dick Son	1909 Shipfitters Shop
Gerald Cain Grandson (Son of Clara Dick Cain)	1932 Planning and Estimating
Michael Cain Great-grandson	1976 Planning and Estimating

Submitted by Sharon Cain, wife of Gerald Cain

John Dick Father	1903 Boiler Shop
Carl Dick Son	1909 Shipfitters Shop
William G. Truemper Grandson (son of Mary Dick Truemper)	1941 Electric Shop & Design
Daniel W. Truemper Great-grandson	1970 Insulator Shop
Jane Truemper Allison Great-granddaughter	1978 Tool Shop
Michael J. Truemper Great-grandson	1980 Pipe Shop

Submitted by Winnie Truemper, wife of William G.

Joseph W. Dilks Father	1904 Boiler Shop
Joseph N. Dilks Son	1914 Pipe Shop
Margaret Dilks Munson* Granddaughter	1941 Supply
Allen J. Munson Great-grandson	1962 Engineer, 260.4

Guy W. Dow Father	1941 Rigger Shop
Eunice Dow Daughter	1941 Tool Shop
Dorothy Reinhardt* Granddaughter	1978 Photo Lab
Cheri C. Zinter Great-granddaughter	1984 Pipe Shop
Kevin Zinter Great-grandson	1988 Apprentice Sheetmetal Shop

Edward E. Edmonds Father	1902 Boiler Shop
John F. Edmonds Son	** Personnel Supervisor
Edward Edmonds* Grandson	** Computer Systems Analyst
John F. Edmonds Jr. Grandson	** P&E Supervisor
Leslie Dillon Great-granddaughter	** Apprentice Machinist
Vickie Cates Great-granddaughter	** Code 130
Michael Edmonds Great-grandson	** Apprentice Pipefitter

John Denom Evans Father	1942 Sheet Metal Shop
Paul G. Evans Son	1941 Pipe Shop
Dale E. Evans* Grandson	1968 Outside Machine Shop
James E. Evans Great-grandson	1982 Outside Machine Shop

Thomas Fellows Father	circa 1904 **
George Fellows Son	1915 Shipfitter
Florence Fellows Levin* Daughter	circa 1925 Pay Office
Walter Fellows Son	** Sheetmetal Progressman
Alice Fellows Lawson Daughter	** Supply Department
Robert Levin Grandson	1954 Apprentice pipefitter
Charles Richard Levin Grandson	1960 Apprentice Electronics Tech
Monte Levin Great-grandson	1989 Shipfitter

Edwin Froggatt Father	1917 Quarterman
Lowell Anderson Son	1937 Crane Operator
Shirley McConnell* Granddaughter	1969 Security Specialist
Tami York Great-granddaughter	1987 Security Assistant

J. L. Halffman Father	1916 Outside Machinist
George Halffman* Son	1932 Outside Machinist Supervisor
George Halffman Jr. Grandson	Present Supervisory Progressman
Brett Halffman Great-grandson	Present Rigger Shop

Hurshell Hall Father	1919 Chief Quarterman Rigger
William E. Hall Son	1941 Welder
Mickey D. Hall* Grandson	1963 Public Works Group Supt.
Kenneth M. Hall Great-grandson	1990 Engineering Co-op

Connie Osborne Mother	1919 Payroll Supervisor
Dorothy Sherman Daughter	circa 1942 Comptroller Department
Charles M. Sherman* Grandson	1951 Coppersmith
Charles M. Sherman Jr. Great Grandson	1974 Engineering Technician
Kimberly Sherman Great-granddaughter	1977 **

George A. Schenck Father	1940 **
Shirley Schenck Rebbe Daughter	** **
Wanda Schenck Karst Daughter	** **
Susan Karst Pace Granddaughter	1973 Engineering Technician
Darrell Krouse Great-grandson	1989 Apprentice Electronics Mechanic
Lissa Krouse Great-granddaughter	1989 Apprentice Painter

Submitted by C. J. Karst, brother of Susan Pace.

Henry Clay Sullivan Father	1924 Pipe Shop
Joe Sullivan Son	1944 Fire Chief
Jack Sullivan* Grandson	1953 Assistant Fire Chief
Joe Timothy Sullivan Great-grandson	1973 Pipe Shop

Charles Weberg Father	1907 Machine Shop
Agil Peterson Son	1920 Electric Shop
Cecil V. "Pete" Peterson Grandson	1953 Pipe Shop
Jim Peterson Great-grandson	1971 Electric Shop

Source: SALUTE October 11, 1974.

George T. Worland Father	1918 Shipfitter Shop
Lawrence D. Worland Son	1940 Sheetmetal Shop
George W. Worland Son	1941 Shipfitter Shop
Teckla Worland Sutherland* Daughter	1943 Code 380
John J. Worland Son	1954 Sheetmetal Shop
Edward Worland Grandson	1965 Planning & Estimating
Dale Coyle Great-grandson	1988 Apprentice, Machine Shop

APPENDIX C

GLOSSARY

ABREAST — Side by side.
ADMIRAL — Highest rank in the Navy.
APPROPRIATION — Congressional authorization to spend government funds for specific purposes.
ASROC — An anti-submarine rocket.
AVAILABILITY — A period in a ship's operating schedule to make her available for accomplishment of work at a repair activity.
BATTERY — Ship's guns of the same caliber or used for the same purpose.
BILGE — Inside bottom of a ship.
BILGE KEELS — Fins at the turn of the bilge to reduce rolling.
BLISTERS — Built-in bulge on the sides of large ships to protect against damage from mines, bombs and torpedos.
BOW — Forward or front part of a ship.
BUILDING WAYS — An inclined framework upon which a ship is built and from which it is launched by sliding into the water.
BULKHEAD — Wall or partition in a ship.
CAISSON — Entry gate into a dry dock.
CHRONOMETER — An accurate navigational clock.
COFFERDAM — Void space between a compartment and the hull of a ship.
CIA AREAS — Controlled Industrial Areas with restricted access.
COLD IRON — A ship condition with unlit boilers and main machinery components secured.
COLLIER — A vessel specifically designed to carry coal.
CON — A short term for control of a ship's movements underway.
CORPSMAN — An enlisted person in the medical corps.
DEGAUSSING — Changing the magnetic field of a ship.
DEPERMING — Reduction of the permanent magnetic field of a ship by energizing coils placed temporarily around the ship.
FARMING OUT — Contracting work outside the Shipyard.
FO'CS'LE (properly spelled "forecastle") — Forward section of the weather deck.
FRIGATE — A large escort ship for a battle group.
GYROCOMPASS — A ship's main compass, which is not influenced by the earth's magnetic field.
HATCH — An access opening in the deck of a ship.
KAMIKAZE — A Japanese suicide airplane.
METROLOGY — The science of weights and measures.
ORDNANCE — A collective term for guns, missiles, torpedos, bombs and related equipment.
PARKERIZING — A process of removing rust from metals with a phospheric acid solution.
PORT — Left side of a ship.
PRI-FLY — Primary flight control center.
STRAKES — Longitudinal steel plates which make up the sides of a ship's hull.
SHEERSTRAKE — The uppermost complete strake at the weather deck line of a ship's hull.
SHORES — Portable wooden beams used in damage control in restraining damaged bulkheads against sea water pressure.
STARBOARD — Right side of a ship.
WHERRY — A light pulling boat.

INDEX

BREMERTON PRINTING CO

ERRATA

LOCATION	SHOULD READ
Pg. 32, Caption	great weight of granite, not great weight of concrete.
Pg. 33, Caption	March 1, 1913, not March 2.
Pg. 45, Caption	Lyle McLeod, not Lyle McCloud.
Pg. 53, Caption	two target rafts, not two small barges.
Pg. 58, Para. 5, Ln.11	until 1948, not until 1954.
Pg. 71, Para. 5, Ln. 8	Roger Paquette, not Roger Pacquette.
Pg. 76, Caption	shop buildings extending, not shops extending.
Pg. 81, Caption	MAHOPAC, not MOHOPAC.
Pg. 85, Caption	Silver Spring Brewery, not Silver Sprint.
Pg. 112, Para. 4, Ln. 6	Engebretson and Kermit, not Engbretson and Kermat.
Pg. 117, Para. 6, Ln. 2	Grandys, not Grandies.
Pg. 132, Caption	YTB, not YTV.
Pg. 135, Para. 1, Ln. 4	Herman Petersen, not Herman Peterson.
Pg. 142, Caption	Connie Osborne, not Connie Osburn.
Pg. 150, Caption	December 6, 1958, not December 7.
Pg. 152, Para. 9, Ln. 3	missile launcher, not missiles launcher.
Pg. 200, Caption	Building 66, Building 109 and the Parkerizing, not Building 66 and the Parkerizing.
Pg. 201, Caption	Building 66, built in 1900, not Building 66, built in 1899.
Pg. 208, Para. 8, Ln. 5	worked on the project as part, not worked as part.
Pg. 244, Note 28	Marine Corps General, not Marine Corps Commandant.
Pg. 245, Note 25	Walter McConnell, Ralph Siegner, not Walter McConnel, Ralph Seigner
Pg. 256, Note 18	Wortman became Superintendent, not became Master.
Pg. 258, Note 21	NAVAL SHIP SYSTEMS COMMAND TECHNICAL NEWS , July 1971, pages 15-20, not 1967 Command History.
Pg. 263, Para. 9, Ln. 4	Bill Biltoft, not Bill Billtoft.

NOTES

NOTES

NOTES

NOTES